NATURE-MADE ECONOMY

Inside Technology

Edited by Wiebe E. Bijker and Rebecca Slayton

A list of books in the series appears at the back of the book.

NATURE-MADE ECONOMY

COD, CAPITAL, AND THE GREAT ECONOMIZATION OF THE OCEAN

KRISTIN ASDAL AND TONE HUSE

The MIT Press
Cambridge, Massachusetts
London, England

© 2023 Massachusetts Institute of Technology

This work is subject to a Creative Commons CC-BY-NC-ND license.

Subject to such license, all rights are reserved.

The MIT Press would like to thank the anonymous peer reviewers who provided comments on drafts of this book. The generous work of academic experts is essential for establishing the authority and quality of our publications. We acknowledge with gratitude the contributions of these otherwise uncredited readers.

This book was set in ITC Stone Serif Std and ITC Stone Sans Std by New Best-set Typesetters Ltd. Printed and bound in the United States of America.

Library of Congress Cataloging-in-Publication Data

Names: Asdal, Kristin, author. | Huse, Tone, author.
Title: Nature-made economy : cod, capital, and the great economization of the ocean / Kristin Asdal and Tone Huse.
Description: Cambridge, Massachusetts : The MIT Press, [2023] | Series: Inside technology | Includes bibliographical references and index.
Identifiers: LCCN 2022038601 (print) | LCCN 2022038602 (ebook) | ISBN 9780262545525 (paperback) | ISBN 9780262374415 (epub) | ISBN 9780262374422 (pdf)
Subjects: LCSH: Atlantic cod fisheries—Economic aspects. | Aquaculture industry.
Classification: LCC SH167.C6 A83 2023 (print) | LCC SH167.C6 (ebook) | DDC 338.3/718—dc23/eng/20221118
LC record available at https://lccn.loc.gov/2022038601
LC ebook record available at https://lccn.loc.gov/2022038602

10 9 8 7 6 5 4 3 2 1

CONTENTS

ACKNOWLEDGMENTS

This book would not have been possible were it not for all the people who have generously shared their time and knowledge with us. Ideas and drafts have been presented to a vast academic community since we started working on the encompassing, collaborative endeavor of this book in 2016, as part of the ERC starting grant project Enacting the Good Economy: Biocapitalization and the Little Tools of Valuation ("Little Tools"). While doing field work and interviews and collecting materials, we have been warmly welcomed all along the Norwegian coast and abroad, been given access to key market sites in the UK, India, and China, and have enjoyed the privilege of learning alongside practitioners who live from and care for the Atlantic cod and its economies.

The work has benefitted greatly from the financial support of the Little Tools project (funded by the European Research Council, 301733), Value Threads (The Research Council of Norway, 301733), and UrbTrans (The Research Council of Norway, 300929; Tromsø Research Foundation, 19_SG_TH).

Our tight Little Tools research team has been our most patient and challenged community, its members having read closely and commented upon multiple versions of the manuscript. Thanks a lot to Bård Hobæk, Silje Morsman, and Tommas Måløy, and especially to Beatrice Cointe and Hilde Reinertsen, who were close readers until the very end. Our larger STS research group at the TIK Centre for Technology, Innovation and Culture has likewise been a steady support and academic stronghold during the whole writing process. In particular, we want to thank Tone Druglitrø, who gave valuable comments on our final draft manuscript, and Stine Engen and Marie Stilling, who have read and insightfully commented on chapters of the final draft version. We also thank Jussi Pedersen, whose insights into the legal frameworks of the ocean and its commons was invaluable to us.

The book has furthermore benefited immensely from the generous ideas, inspiration, challenges, and comments by participants at a range of seminars organized by our Little Tools research group in Tromsø, Lofoten, Tøyen, and Hamarøy and at TIK. Our seminars at sometimes distant places and where participants had often traveled far to take part, gave space to discuss drafts, often at a very early stage, in lively, fun, and thoughtful academic surroundings. Thank you so much to Alexander Dobeson, Ana Delgado, Andrew Barry, Ask Greve Jørgensen, Berit Kristoffersen, Brigt Dale, Camilla Brattland, Eve Chiapello, Hannah Knox, Jahn Petter Johnsen, John Law, Kaushik Sunder Rajan, Kregg Hetherington, Liliana Doganova, Marit Ruge Bjærke, Narve Fulsås, Petter Holm, Pierre Delvenne, Susanne Bauer, Svein Atle Skålevåg, and Terje Finstad. Particular thanks to Eve Chiapello, who challenged and inspired the direction of the book at a very important point in the writing of the manuscript.

While finishing work on the manuscript, Kristin was a guest professor at the Centre de Sociologie de l'Innovation (CSI), Mines, Paris, which provided the ideal combination of a calm and inspiring intellectual environment. A warm thanks to colleagues at the CSI for their wonderful seminar culture and their readings and commenting on selected chapters of the manuscript—Madeleine Akrich, Liliana Doganova, Fabian Muniesa, Vololona Rabeharisoa, Kewan Mertens, Morgan Meyer, Brice Laurent, Beatrice Cointe, Alexandre Mallard, Jérôme Denis, David Pontille, and Roman Solé-Pomies. A special thanks to Liliana, who made space for discussions and exchanges about the book as part of our collaborative Value Threads project.

The Norwegian College of Fishery Science at UiT The Arctic University of Norway welcomed Tone as a guest researcher during the autumn of 2017, providing her with a vibrant and interdisciplinary research environment dedicated to the matters of the ocean and its politics. A special thanks here goes to the MARA research group for an insightful and engaged reading of early text drafts, and especially to Camilla Brattland, Maike Knol-Kaufmann, Jahn Petter Johnsen, Peter Arbo, and Petter Holm. Tone has further benefited greatly from the support of the UrbTrans research team, Anna Andersen, Anna Jensine Arntzen, Martin Svingen Refseth, and

Prashanti Mayfield, as well as from her colleagues at the Department of Archaeology, History, and Religious Studies at UiT The Arctic University of Norway.

At the University of Oslo, we would very much like to thank the broader research environment that shares our interest in the cod fish, its mysteries, and its ways of living. Thank you to Kjetill Sigurd Jakobsen, in particular, who has provided space for collaboration, social science, and cultural studies in the broader research group at the Centre for Ecological and Evolutionary Synthesis (CEES). The support from the Aquagenome project and the later COMPARE project and research group has been and continues to be an inspiration for interdisciplinary research. Thank you to Shou-Wang Qiao, Sissel Jentoft, Finn Eirik Johansen, Naomi Croft Guslund, Monica Hongrø Solbakken, Alexandra Jonsson, Helle Tessand Baalsrud, and Adrian Lopez Porras.

In doing the final writing of and revisions to the manuscript, two groups of people have been invaluable. First, our wonderful scientific assistants, Eli Sandmo Brenna, Sofie Nebdal, and Ty Tarnowski. Thank you! Second, but not least, our reviewers. They could not have been more helpful. And here we need in particular to thank John Law. His support has been massive and amazing. Without him, this would have been a different book. We dedicate this book to him, and want to express our gratitude for John's prolonged interest in and generous contributions to our ideas, texts, and projects. Thank you to the MIT Press and to Katie Helke who welcomed and directed us so professionally and neatly into the Inside Technology book series and for choosing our important reviewers and readers.

The backbone to everything comes last. At home we have been so fortunate to have enthusiastic and knowledgeable readers, whose input in the more frustrating hours of writing has been invaluable in making this book move forward and become what it is. Thank you to Oddgeir Osland and Torgeir Knag Fylkesnes!

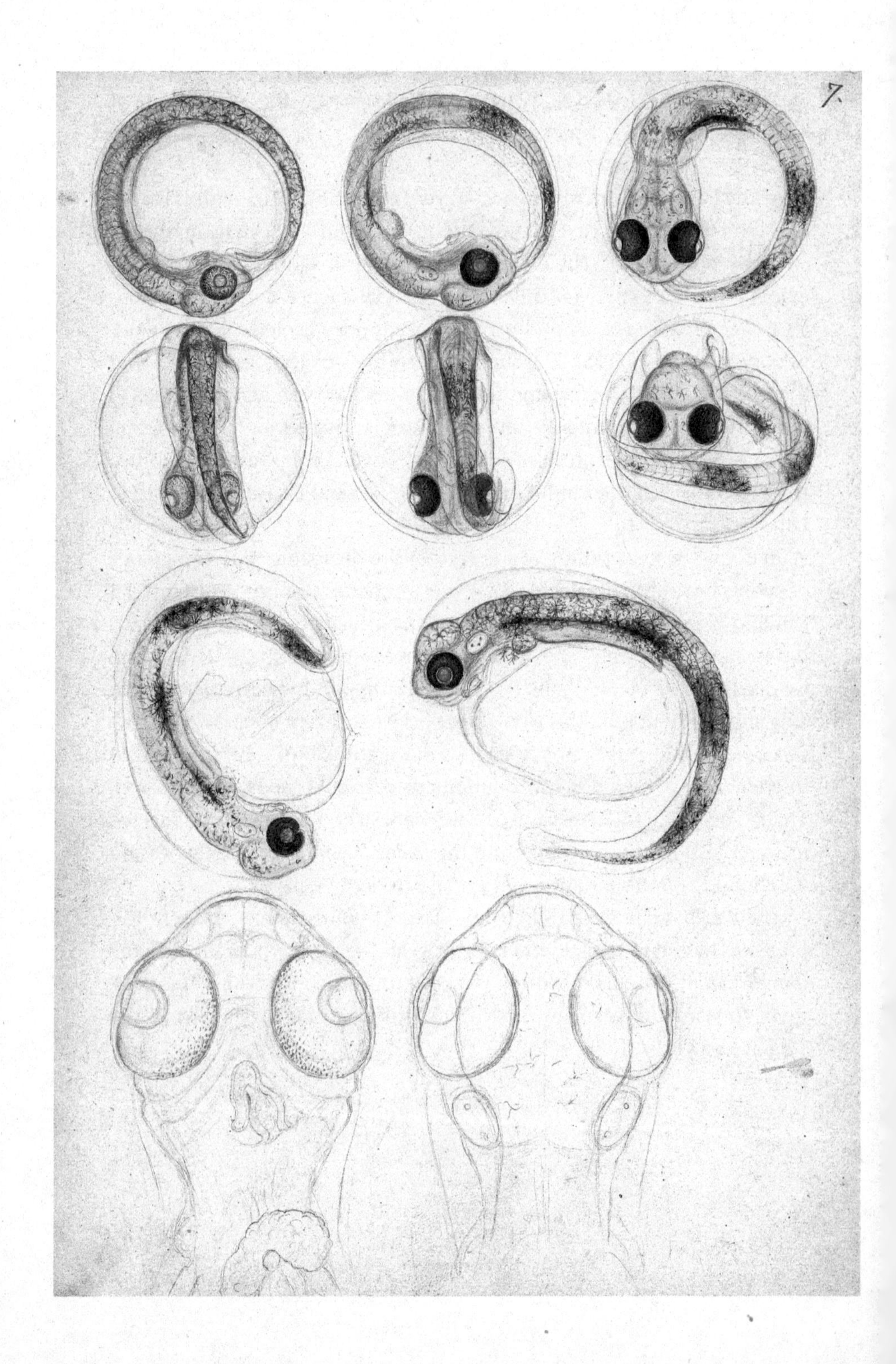
7.

INTRODUCTION: THE GREAT ECONOMIZATION OF THE OCEAN

Far, far to the north, above the Arctic Circle, its southernmost borders hemmed in by the lands of Norway and Russia, lies the Barents Sea. It is a place of extremes, a liminal liquid territory where in summer the sun never sets, while in winter it never appears. Seen from the metropolises of the world, it is a space that stretches beyond the margins of civilization, an unruly place not easily mastered by humans. For the northeast Arctic cod—a cod fish that also goes under the Norwegian name *skrei*—the Barents Sea is, however, the perfect place of dwelling. Always cold and teeming with prey, it is home to hundreds of millions of this temperature-sensitive fish. Here, the now largest remaining cod stock in the world spends the cycle of the seasons in vertical migration between warmer and colder layers of the ocean, going deep down when the nutritious surface gets either too hot or too cold. When its reproductive hormones kick in, however, so does the cod's wanderlust: In the dark of winter, the sexually mature cod will venture on a journey as long as maybe 1,000 kilometers, from the Barents Sea to its breeding grounds along the Norwegian coast (Marinbiologene DA, n.d.). It is this journey that inspires the Norwegian denomination *skrei*, which stems from the Norse word *skreið* and is not so unlike the English word "stride." The name *skrei* is also what sets this great "strider" apart from its coastal kin, the fjord cod, which often belongs to specific fjord systems and which spawns locally.

Together with the cod swimming in the northern Atlantic Ocean, from North Carolina to Kalaallit Nuunat, off the coast of Iceland, and along the coasts of Europe, from the Bay of Biscay to the Barents Sea, the cod that roam the oceans and fjords of Norway make up the Atlantic cod—the *Gadus morhua*. Over the last fifty years or so, however, this species of fish has undergone quite dramatic changes. Many of the great cod stocks of the world have been overfished to the point of being near extinction, but the cod family is also being added to, through efforts to breed and farm the Atlantic cod. Hatched in the laboratory, bred in onshore tanks, and reared in net pens in the sea, these domesticated cod are the result of about seventy years of scientific experimentation and research. The dream to cultivate the cod is, however, at least twice as old.

The Atlantic cod, the fish that people along the Norwegian coast like to call "the white gold" of the ocean, is the main character of this book. It is the species and creature with whom we have chosen to travel and think, to learn from, and enact as we explore what we denote *the great economization of the ocean*; an economic transformation in which the ocean, its environments, and beings are set to be exploited at all depths and surfaces, and to an unprecedented extent and magnitude. The very status of nature is with this transformed, an issue that is also raised in by the very work that our notion of the ocean's great economization paraphrases, a classical work in political economy, *The Great Transformation: The Political and Economic Origins of Our Time* by Karl Polanyi ([1944] 2001).

Published just as World War II was about to end, and so two decades before the events our book examines, *The Great Transformation* describes the rise of the market economy in the eighteenth and nineteenth centuries and delivers a stark critique of what Polanyi saw as the cataclysmic effects of capitalism on human societies. The book also emphasizes the role played by the modern state in transforming diverse traditional economies into what Polanyi perceived as a "market society," and points to the commodification of land and nature as key to this transformation. Whereas land had earlier been "tied up with the organizations of kinship, neighborhood, craft, and creed—with tribe and temple, village, guild, and church," Polanyi ([1944] 2001, 187) argues, the growing stronghold of capitalism worked to disconnect land from such institutions. Instead, they were subordinated to what he calls "the market mechanism." Polanyi makes this argument through a fine-grained historical analysis, showing how land—understood both as "the soil" and "nature"—was drawn into a market economy and ceased to be, in his words, a place "where labor forms part of life, land remains part of nature, life and nature form an articulate whole" (Polanyi [1944] 2001, 187).

This can be read as Polanyi lamenting the commodification of land, which he does, and vehemently at that, but we also read his work as containing another, less ferociously performed critique. In the words of Polanyi, the subordination of land to a market economy worked to sever the close relations that previously existed between labor, life, land, and nature. Subdued to economy alone, nature was thereby severed from a complex set of relations. In reading Polanyi, one understands that such previous relations should by no means be romanticized, the feudal economies replaced by industrial capitalism having possessed a brutality of their own. It is interesting to see, however, how in Polanyi's analysis, the relationship between nature and economy becomes both strongly hierarchical and homogenous; it is reduced into being one.

What this book argues, however, is quite the opposite. Looking not to the land but to the ocean, we find that as its economic exploitation intensifies, such nature–economy relations multiply accordingly. The relationship between nature and economy intensifies *and* diversifies. Consequently, what we observe is not *one* economy, but instead what we conceptualize as *versions of economization*. Working through a broad set of empirical cases and materials, we identify such versions of economization as they emerge. We focus, moreover, on economizations that are intertwined with both the cod in "the wild" and the cod domesticated, and we follow these to several, quite different sites, out at sea, in laboratories and net pens, as part of government bodies and research papers, and in markets.[1] With that, our take on the great economization takes us to the very practices of its making—of constituting living entities of nature as objects of economy, of making them economic, thereby also transforming the ocean and the societies they land in and work on.

In addressing the great economization of the ocean, our point of departure is not how this unfolds today, in contemporary practices. Rather, the book starts out by examining how such economizations have emerged, been enabled, and played out in the past. Chapter by chapter we show the conditions of possibility that current visions and programs rest on, and how what we today can recognize a great economization builds on long-standing efforts, experiments, and laborious investments, but also tense controversies, environmental troubles, mass losses of life in the ocean, and other quite spectacular failings. Opening with two distinct versions of ocean economization—the partitioning of the ocean into "exclusive economic zones" and the division of the Norwegian seabed into blocks for petroleum drilling—we investigate this new ordering of the ocean and the controversies that followed in its wake. We analyze the version of economization that seeks to transform the Atlantic cod from being a "wild" species that moves by its own instincts to becoming a species of biocapital and a farmed animal. And we show how the farming of fish was enrolled in two, radically different and competing versions of economization. One envisioned fish farming as a form of locally based culture, the other pushed to use fish farming as an entry point for growing the ocean and growing big. We demonstrate the coming into being of the version of economization that we have come to be familiar with as an economy of innovation, and we follow economization practices linked up to innovation as efforts are made not only to grow the ocean, but also global markets. Sometimes we delineate versions of economization that are quite small and local, for instance, a fish landing station's decision to hang and dry the cod instead of selling it as a fresh cod commodity. Other versions of economization described in the book amount to larger paradigms, come with their own set of theories, and even take on names of their own.

Where do economizations happen? They take place not only out at sea, this book shows, but also in markets, in the oft-connected spaces of research and entrepreneurship, and, not the least, within the bodies of the state. The Atlantic cod, in turn, embodies these sites in very different ways. In policy documents it moves as a resource to be managed and grown; in experimental science it becomes a being to rear and train; in market surveys it is a piece of flesh with qualities to promote and name. In some instances, the cod ceases to be a fish altogether and is instead enacted as "the biology." It becomes a form of biocapital, a stock, or a commodity. It is drawn into the strategies and plans of variously named economies, such as "the ocean economy" and "the bioeconomy," or it is made the object of "blue growth" and "innovation." In the great economization of the ocean, many versions of the Atlantic cod are at play.

In the pushes for economization, we find, it is not one transformative "force" that is at work. Economizations do not work by one, overriding logic or by predetermined market mechanisms. What this book seeks to capture is therefore not some inner logic or system behind the ocean's transformation. Nor is this a book that will inform large economic categorizations, like "neoliberalism" or "capitalism." In fact, we avoid such terms, as we find that they are too predetermined for what is our purpose here. They simply hold too much meaning already, and thereby prevent us from pursuing the type of empirically grounded analysis of economy that we are after. The ambition is to pursue what we suggest thinking of as empirical-near theory building. In other words, to do *empirical economy*—a notion that we return to discuss in chapter 8, but which this book can also be read as a demonstration of. With this, we seek to contribute to opening the study of economy for cultural and social studies, in parallel ways to how science studies for many years have been approaching the natural sciences. In doing this, we do not walk alone, but draw from and develop resources and vocabularies inspired by the proliferate and related fields of social studies of markets and valuation studies. Both these research fields, which we introduce and discuss further in chapter 1, engage closely with the empirical study of economics and economic practices, as well as with developing theoretical and methodological approaches to these as objects of study. At their best they work as a counter-weight to abstract modeling and a vocabulary that serve to alienate students and scholars outside economics from a broader engagement with economy.

Economics and economic practices are sometimes portrayed as being all about the calculative, the quantitative, and the instrumental. A large body of literature has, for instance, critiqued economics for its model of "economic man," or *homo economicus*; a sovereign, rational actor who makes choices only in accordance with his or her narrow economic interest. Rather than pursuing such critique *of* theory and *in* theory, we suggest pursuing an analysis that both complexifies and "empiricalizes" economy. Approaching, for instance, how prices are set, how the value of a species is assessed, how markets come into being, or how economic growth and value creation are manufactured and envisioned, we immediately see also the qualitative, the material, the semiotic, and the normative. We need empirical and conceptual space for all these dimensions to economy, and therefore also a vocabulary that transgresses categories fit for calculation and spreadsheets and that stays attuned also to the elusive, the messy, and the qualitative. That we choose to engage closely with social studies of markets and valuation studies is then motivated by how these fields are oriented toward handling the qualitative as well as the quantitative, and toward engaging with these different dimensions of economy simultaneously. In engaging with and making use of the studies and literatures these two fields offer, we furthermore take on Bruno Latour's (1996) argument that the sciences are instrumentalized; they are equipped with instruments that shape their ways of seeing as well as their ways of moving and intervening. Likewise, we follow economic practices and the instruments they work by, but we also equip ourselves with a vocabulary that can capture these practices beyond their empirical complexities. The book's notions *little tools of valuation*, *valuation arrangement*, and *value orderings* are developed for these purposes. They are set to move across the material and the semiotic, across the quantitative and the qualitative, across markets and state bodies, and then lastly, across nature and the economy.

The notion of little tools ties in with how we, in tracing the movement of cod bodies—from the ocean and into market apparatuses, from research papers and into laboratories, into net pens, and into Parliament—find that these move not on their own. They move in the company of a myriad of little tools, tools that in and of themselves are quite mundane but, when tied in with larger apparatuses and the valuation arrangements of the political as well as of markets, become key to the valuations and orderings that economic practice entails. Furthermore, and as underlined by our designation of these as "little" tools, the notion is intended to operate at an empirical and methodological scale that is close to that of everyday practice and the empirical materials we investigate. "Little" should therefore not be mistaken for "insignificant," but be taken as a designation of the scale at which the material-semiotic entities and artifacts we examine operate. Another key concept and one that speaks quite directly to the title and overriding concern of this book—how nature makes economy—is the notion of co-modification.

THE CO-MODIFICATION OF NATURE AND ECONOMY

The idea that nature modifies economy poses a rather significant challenge to conventional economic thinking. For whether cast as a resource, stock, raw material, or property, nature remains passive. This in turn means that concepts such as "capital," "surplus value," "markets," or "commodity" have little or no capacity to take on what that nature does when made economic and enrolled in economic practices. Their conceptualization and inner logics simply hold no space for the agency of nature, and so they are not equipped to capture what nature does to economy. A main project of the book is to bring the affordances and propensities of nature into conversation with the concepts and concerns raised in studies of economy. And further, to carve out an analytical space for how nature not only is made economic, but also acts within and upon economic practices, often also modifying them. In doing so, we bring social studies of markets and valuation studies into conversation with other bodies of research and theory. For whereas social studies of markets and valuation studies have already done much to rematerialize economic analysis, they have done less to bring the rich and lively worlds of nature and its actions into the analysis of, for instance, capital, markets, and economization. Works within actor-network theory (ANT) and science and technology studies (STS) have on their side been instrumental in bringing nature into other fields of research, creating the type of opening that this book seeks for the study of nature–economy relations. As a move toward this, the chapters of the book bring the literatures of social studies of markets and valuation studies into conversation with extensive empirical cases that speak to how nature acts within and upon economy, as well as with insights and sentiments in environmental humanities, human–animal studies, critical bioeconomy studies, and work on the politics of nature in science and technology studies.

To equip us with the capacity to better observe and take nature into account, we work with and develop the notions *co-modification* and *biocapitalization*. An important starting point for this quest and for our movement with the cod and the versions of economization that it is being drawn into is the notion of co-modification. As originally formulated (Asdal 2015b), this notion speaks to commodification, but with a twist: it alerts us to the practices and work of modifying not only biological entities for commercial purposes, but also markets. Thereby describing a double entendre, the notion of co-modification implies that to turn biological entities into commodities is not a pre-given, linear, or uniquely social process. Instead, it is something we need to explore more openly and with a sensitivity toward how markets too are modified to accommodate the liveliness of biological entities. In this book, we use the notion of co-modification to cast a somewhat wider net, as we examine more broadly the co-modification of nature and economy. Here, the ocean in its various formations, depths, surfaces, and environments is part of the "nature" being modified, along with the cod, an entity that is indeed biological but also so much more. "Economy" is similarly taken to encompass more than markets and market work, as we also move to explore in this book the modifications of production and capital. The sentiment and sensitivities of our analyses, however, are much the same as in the original coinage of co-modification, as we seek to trace these openly and empirically, and always with an eye for what nature does when it is made economic. It is by way of this analytical strategy that we have come to know the cod *not* as a passive puppet of economization, whichever version, but as a being that can act on and sometimes even shape economic practice. By way of its propensities and affordances (Gibson [1979] 2014; see also Hutchby 2001)—its capacity to grow and reproduce, but also to act and react in specific ways—the cod can both lend itself to and resist efforts to enroll it in economic practices. The ocean, a vastly complex and largely unknown place of multiple environments, has proven equally unpredictable and tricky to control. Consequently, a key argument developed in this book is that entities of nature do not enter economy passively, their affordances and propensities being decisive to the types of economic action spun around them.

MULTI-SITED AND MULTI-TIMED

The research of this book moves across many and qualitatively different sites. These are sites where cod "in the flesh" can be seen, held, and smelled—places that also we, as researchers, can embody and move about—but just as importantly, we have pursued a rather different type of site, namely, that of documents. So, while we invite the reader to the "wet" sites of cod habitats—out on the ocean on a beautiful Arctic winter morning or peering into tanks of laboratory-born cod fry—the book also introduces as cod sites the "dry" pages of research papers, policy reports, and innovation strategies, finding within these not the masks of a fishing net or the workings of a fish factory, but other tools and apparatuses set to do economic work. Crucially, and contrary to reading documents as simply representing or carrying the meaning of their author(s), we take the two site typologies of place and paper to be equally rich social realities. In fact, and due to their material mobility, documents are particularly apt at moving realities around and into new contexts. We therefore consider documents and document circuitries to be sites and geographies which we move in and across in our tracings of the cod and its various embodiments, whether in the form of documents copied from the archives of the Norwegian Parliament and ministries or retrieved from the websites of research institutes, corporations, or various public agencies. In doing these moves, we have pursued the method of practice-oriented document analysis (Asdal 2015, Asdal and Reinertsen 2022) to examine how documents do not simply represent a reality "out there," but enact specific versions of it and, alongside this, perform valuations and orderings of their own. By way of these combinations, our analysis is extended, from the here and now of the ethnographic moment to also accounting for past events. Tracing the cod not only in the realities of ethnography but also in those of documents, our study is not only multi-sited but also multi-timed.

As we move and between physical places and spaces—some of the sites the book examines belong to government, others are a marketplace, a research institute, or a fish landing station—we also differentiate between what we suggest thinking of as different *document species*. The notion is meant to alert us to the fact that documents are not all the same, but often belong within a specific genre, like that of a business plan or a policy document. Like no place or space, be it the interior of a fish landing station or the wide-open ocean, is disconnected, neither are documents. They are connected to the wider material-semiotic apparatuses that they are produced within and act on, and with this to worlds or arrangements of their own. As part of, for instance, the apparatus of the state, they can move and circulate; they can attach themselves to issues and change them; they become constituted by a broad array of relations and can never, like the ethnographic site, be captured as what they fully, truly are. For as John Law (2010b; see also Savage 2013; Law and Ruppert 2013) so precisely argues, there is a double social life to our research methods. They are tools we use to examine and understand social reality, but as we put them into use and begin to assemble our descriptions of said realities, we also, and inadvertently, enact distinct realities of our own. As we "trace" the movements of cod, we enact a geography that is ours only, connecting some sites and actors, while leaving others out.

The above is also true for the aspect of our methodology that we describe as multi-timed. On the one hand, we relate to multiple times as being those of past, present, and future. This is a rather homogenous way of thinking of time, as by describing events and developments we, in our presentation of these moments in time, also reenact the well-worn temporality of linear time. On the other hand, with an eye for how time is indeed multiple and diverse, and how temporalities may therefore also become incongruent to one another, time enters our analysis by way of the empirical material. For instance, we examine how the temporalities of cod domestication experiments were coordinated with that of the spring flowering of the ocean, how the timing of the sexual maturation of the cod in captivity came to be at odds with the timings of markets, or how fresh cod commodities' propensity to decay directs the speed of both their production and consumption.

The geography of our research stretches from Vardø in the northeast of Norway and Flødevigen in the very south to the United Kingdom, India, and China. We have visited abandoned fish factories, but also fish landing stations bustling with activity. We have joined a lone fisherman on his voyage out to sea and entered the enormous hull of a trawler, to walk among (or more correctly, upon) the many tons of frozen fresh cod resting inside it. We have witnessed the breeding and rearing of cod fry in the laboratories of public as well as private enterprises and joined a fish tender in feeding cod being reared in a net pen. We have sat among the audiences of seafood conferences and summits, and we have participated in spectacular events organized to promote the cod to international buyers. We have moved with documents as these move issues in and out of government bodies, attaching themselves to cases and laws, but also to economies in the making. In the process of it all, we have also eaten quite a lot of cod, whether served by an Indian chef at a five-star Kolkata hotel or prepared in the ways of Norwegian tradition, like the dish *mølje* consisting of not only the cod's meat, but also its liver and roe. The eating of the cod was, perhaps, not of any empirical importance, but it has been a political act of sorts, in that alongside our tracings of the cod, its politics, and economy, we have also preserved a type of affinity to this fish, which frequented the dinner tables of our childhood and continues to be of high cultural value to us.

LIFE, LOVE, AND LOTS OF TROUBLE

Studying the great economization of the ocean through the specific lens of the Atlantic cod instills in the analysis what we find is a much-needed *perspective*: that of life in the ocean. For as the ocean is now being drawn into political and economic strategies geared toward projects of comprehensive industrialization, the conditions, value, and, ultimately, quality of such "life" is about to be radically transformed. The life of cod is, moreover, not without a significance of its own. It is the carrier of one of the oldest known ocean economies, has throughout centuries enabled the accumulation of great wealth, and has many a time proved itself to be a source of both struggle and survival along Norway's long and rugged coast. It is a meaty matter, both in that its bright, white flesh is highly desired and in that it is a matter of consequence, its relations to the world being both rich and substantial.

In many respects, the cod can be seen to represent an iconic case of capitalizing on the ocean. As put by Mark Kurlansky (1997) in his biography of the cod, it is "the fish that changed the world." Along the Norwegian coast, the coming of the cod, in great numbers and close to shore, has been a source of survival and wealth since time immemorial. The cod enabled the Vikings' intercontinental travels by sea, the Vikings surviving the journey on dried cod meat. The finding of a thousand-year-old cod bone in Hedeby, Germany, further indicates that it was one of the first commodities ever to have been exported from Norway (Star et al. 2017). It was also one of the first major protein foods to be extensively traded internationally and came to represent an important source of nutrition in medieval Europe, but also, and in a far darker chapter of the cod trade, in the slave colonies of the Caribbean (Kurlansky 1997). By the Middle Ages, cod exports to Europe had also become key to the national economy of Norway. Not only personal fortunes but cities were being built on the back of its profits, and so struggles to control the cod resource surged. As historical works reflect, conflict over the right to fish, tax, and trade the cod is as old as the cod economy itself (Døssland 2014; Hutchinson 2014; Kolle et al. 2017).

Then as now, the most important cod stock of the Norwegian fisheries was the northeast Arctic cod of the Barents Sea—the *skrei*. The *skrei* seek out breeding grounds as far south as along the coast of Møre, but they have been particularly attracted to the grounds outside the Lofoten archipelago. Here, from January to April, what has been described as "the greatest love fest of spring" takes place: the female cod release an enormous number of eggs, the males deliver load upon load of the fertilizing milt, and fishing is good. Tons of big, fat cod are pulled out of the sea, landed along the coast, and shipped off to markets across the world. Still, as those who depend on the cod for a living know very well, the cod can also be a fish that offers trouble and resistance. In some years, the great numbers of cod will simply fail to appear at their usual breeding grounds, the ones that do show up being thin and in poor shape. The unpredictability of cod in "the wild" has, in turn, been one of the main motivations for domesticating and turning it into a farmed species—these are aspirations that date back at least a hundred years but gained force in the late 1960s. Removed from their natural element not by the hook of a fishing line, but by science and breeding, the genes of the *skrei* are crossed with those of the fjord cod. The aim has been to create a fish that conforms to life in captivity, grows fast, and is resistant to disease. Still, the outcome of also this particular "love fest" has proved to be quite difficult to control. As we explore in several of the book's chapters, the cod has resisted life in captivity by more than one means: escape, early sexual maturation, and cannibalism being only a few of them. In numbers, the farmed cod are today still a modest stock, but these new members of the cod family are still quite important. They embody decades of research and investment in the fish farming industry, hold the hope of future economic growth, and have interesting stories to tell about the ocean. For alongside the domestication of ocean beings, we find also that the ocean is sought to be transformed. It is being sculpted into becoming more like a farmland, a place to be cultivated and set up for production, a site for multiplying and rearing the living. Consequently, while we underline the importance of attending to the mundane and the "little," we also wish to hold on to our "big" notion of a great economization. It is a necessary reminder of the scale and consequence of change that the ocean and its beings have been, are, and most likely will be experiencing as their economic exploitation continues—an exploitation, we would also like to stress, that takes place amid enormous environmental stress caused by decades of pollution and overfishing and by the still unforeseeable consequences of global climate change.

So, is this book about the economization of the cod or of the ocean? There is already a not so huge, but beautiful social, economic, and cultural studies literature on the ocean, which our book speaks to and has taken inspiration from. To tell *our* story of the great economization of the ocean, the cod fish is the critical case that we work from. Being one of the ocean's most important and critical agents and species, economized for so long, in such manyfold ways, and across vast geographies, it can tell us a lot about the economization not only of a species, but of the ocean and, crucially, the state. It is a species that enables us to address how nature made and makes economy. The cod drawings by the zoologist Georg Ossian Sars (1837–1927), which travel with us throughout this book and open each of its chapters, underline exactly this point. Sars's drawings were intimately connected to his investigations of the ocean, commissioned by Parliament and a state interested in understanding and strengthening its ocean economy. The story they are part of is itself a case of how nature made economy. Sars's careful, tender, and amazingly detailed drawings are simultaneously strikingly evoking and alert us to how nature and economy elude a straightforward and simplified calculative and quantitative framing. The re-presentations of the cod through Sars's close-to-photographic depictions evoke worth, the appreciation of a species, the valuation of life coming into being and growing. They are at the same time observations of the unique and vulnerable, and of belonging to a species and an ocean environment. As the cod fish grows, from the tiny egg and larvae to a large fish, itself ripe with roe, the economy grows too.

The chapters of this book can be read in this way; chronologically. As the cod fish grows, thrives, and matures, so does economy. Yet, each chapter also tells its own story, deals with its own trouble, problems, and failures. As such they can be taken out of their linear and tamed history and read for their own journey and analysis:

Chapter 1 lays out the theoretical and analytical resources that we build on, expand, and add to throughout the book, including a glossary summarizing the conceptual vocabulary that the book introduces. Chapter 2 concerns the radical reordering and revaluing of the ocean that took place in the decades following World War II and the introduction of offshore drilling for petroleum. It explores the conflicts that arise in what is an increasingly crowded ocean commons, with now several versions of economization overlapping, their value orderings coming into tension and conflict. Chapter 3 moves to consider the work done on cod bodies in the 1960s and 1970s to domesticate the species and make it a fish apt for industrial farming. It introduces the notion of biocapitalization to describe how stocks of cod are raised and reared to take on accumulative properties, emphasizing the co-modifications that take place in efforts to constitute biocapital. Chapter 4 looks to the tensions of value orderings in conflict as the domestication of the ocean proceeds in the 1980s—one such value ordering calling for growth in par with care for the living, local ownership, and modest ocean cultivation; another pushing for ocean growing on a large scale, but also enacting a nature imbued with carrying capacities and thresholds, and that warrants consideration. Chapter 5 shows how the innovation economy comes into prominence as an economy where failure is an opportunity for learning, dynamism and collaboration are key traits, and where the cod is enrolled as a form of biocapital in a fish farming industry set to be "revolutionary," but that ultimately fails spectacularly. Concentrating on fresh, ocean-captured cod commodities, Chapter 6 brings out how prices can perform tasks and carry valuations far beyond those of markets, and instead be what we conceptualize as prices-for-collective-concerns and more-than-market agencies. The chapter furthermore challenges the idea that commodities are objects rendered passive, and brings out the ways in which cod fish, also after their death, co-modify production and exchange. Chapter 7 follows the cod to China and into market architectures and designs here being made to capture a space for it in the growing domestic consumer market. It traces this work across the hybrid valuation arrangement of the Seafood Council of Norway, finding that market work can grow not only markets and market shares, but also the extension and capacities of the state. Chapter 8 concludes the book, drawing together the contributions of the chapters to argue for an empirically attuned study of economy that can allow for new methodological and conceptual takes on this field.

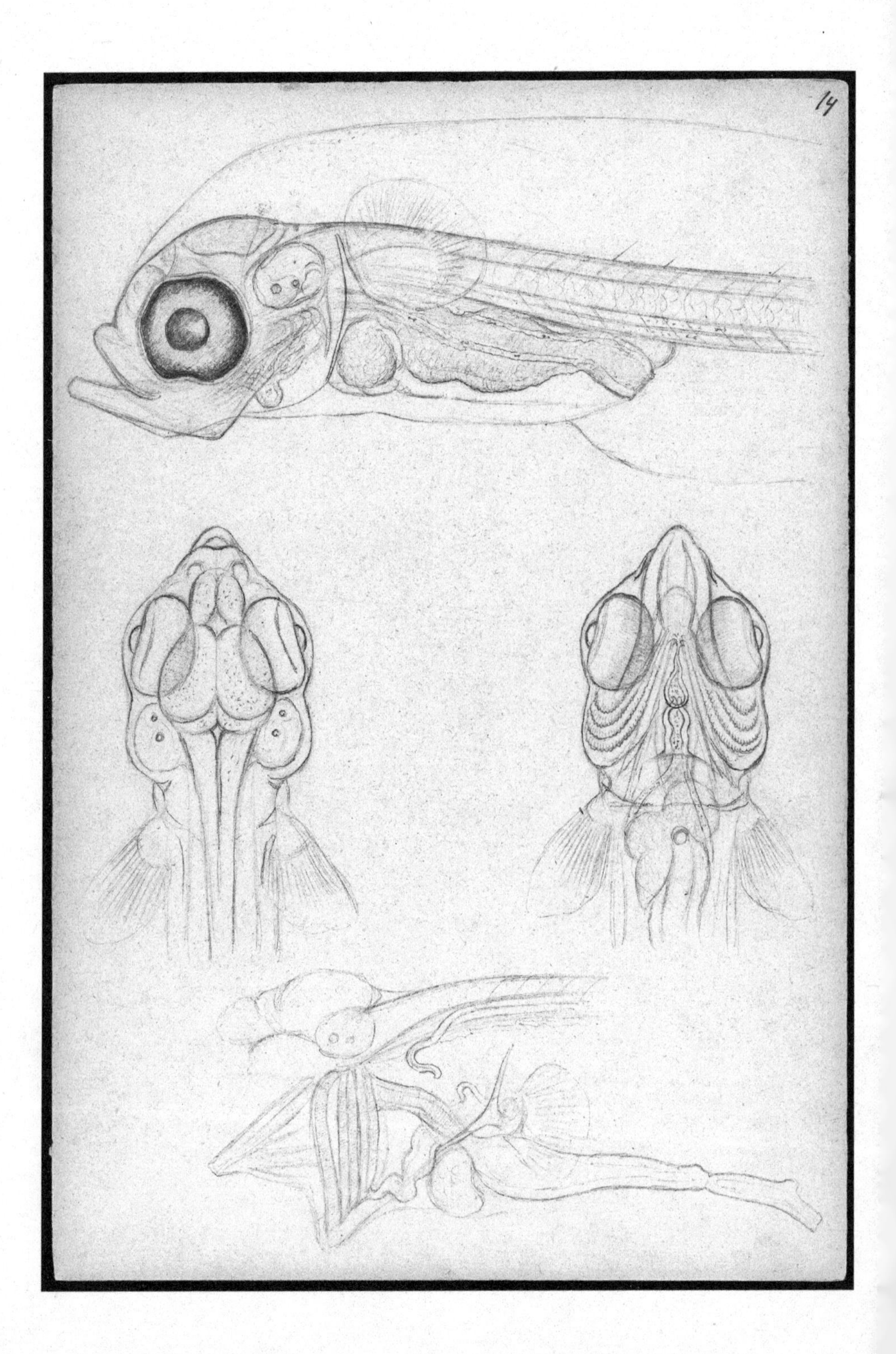
14

1 WRITING NATURE ECONOMY RELATIONS

As this book moves forward to analyze the cod and its entanglements in a rapidly transforming ocean economy, we need both an analytical apparatus and an empirical sensitivity that is tuned toward bringing together the social worlds of markets and the economy *and* the natural world of species' liveliness and affordances. Only then can we begin to grasp the ongoing economization of the ocean and the struggles and troubles that this involves. Put even more ambitiously, only then can we hope to grasp how economy is constituted on nature, how nature is made part of economic practices, and how it also reacts and modifies economies. The scholarly approach we are after is therefore one that provides space for assessing not only how the living is turned into commodities, but also how the living—the Atlantic cod, in our case—reacts, resists, and troubles processes and practices of commodification and market making.

Just as much as this book follows the Atlantic cod, traces its movements along with others who seek to act on it, and takes the ocean as its broader "landscape," it also travels together with some entirely different objects and collections, namely, a multitude of books, articles, and references. The book swims, in a manner of speaking, in an intellectual or scholarly field of words and accounts—and these are situated, too. In finding our ways through this world of literature, we have engaged closely with what can be described as a lack of conversation between two burgeoning research fields that are both linked up with the interdisciplinary field of science and technology studies (STS): On the one hand, is the rich repository of studies in science and technology studies and beyond—in anthropology, feminist technoscience, and environmental humanities—that has worked to bring out the affordances and agencies of nature and the nonhuman (e.g., Haraway 1988, 1989, 2003; Latour 2004; Bingham and Hinchliffe 2008; Friese 2015; Despret 2016; Dooren, Kirksey, and Münster 2016; Asdal, Druglitrø, and Hinchcliffe 2017; Schlünder 2017; Druglitrø 2018; Lien, Swanson, and Ween 2018; Svendsen 2020). This is research field has done a rather massive job of bringing out ways in which the nonhuman can act with or without, on, and alongside the human. On the other hand, there is the growing body of literature that in intriguing and important ways is opening up "the economy" to empirical research, emphasizing how things are rendered economic and the materialities and devices involved in such processes. This has inspired and reinvigorated research fields under the names of "valuation studies" and "social studies of markets." Still, this strand of research has to a large extent treated nature as passive and the nonhuman as entities that do little to intervene in and modify how economy is manufactured and practiced. Put bluntly, what the one field of research animates, the other renders inert. This is a problem not only when one seeks to study the cod and the various efforts made to capitalize on its bright white and tasty flesh, but in studies of nature economies more generally.

How to study the economy without simultaneously pacifying nature? How to avoid treating the economy as simply a force, a backdrop, or logic from which nature cannot escape? How to move nature "back in" to the study of market making, commodity production, innovation schemes, and economic policy programs? Put differently, how can we bring the theories and analytical apparatuses by which the economy is being studied "down to earth" so that we rematerialize, but also *renaturalize*, the study of economy? In seeking to answer these research questions and methodological challenges, we take our point of departure in the closely related, and often overlapping fields of valuation studies and social studies of markets, yet let scholarly contributions for exploring nature and nonhuman agency assist us in intervening in these fields. In doing this, we partake in efforts to develop means by which an active nature can be brought into the study of economy, but also to bring in tools and methods by which the economy can be opened up for cultural and social studies, with the same empirical curiosity as is granted social and cultural studies of nature.

In the following we will introduce the two field of valuation studies and social studies of markets with a focus on the research questions that have guided and directed them. We will then move on to the conceptual takes and approaches that we find equips us to handle our own research questions. We will show how the question of value can be re-composed toward our own purpose of bringing studies of nature and economy together, and more broadly take part in our quest toward empirical economy that is both re-materialized, but also re-naturalized.

BRINGING THE STUDY OF ECONOMY DOWN-TO-EARTH

Valuation studies and social studies of markets are two distinct, yet closely related and partly overlapping research fields. They include a wide range of scholars working across accounting, marketing, organization theory, anthropology, geography, sociology, history, and science and technology studies (see Frankel, Ossandón, and Pallesen 2019; Asdal, Doganova, and Fochler, forthcoming). Three streams of literature have been and continue to be particularly influential. One of these originates from what is labeled the "New Economic Sociology" and that developed predominantly in the United States since the mid-1980s (Frankel, Ossandón, and Pallesen 2019). Key examples are the works by Neil Fligstein (1996), Mark Granovetter (1985), the seminal contributions by Viviana Zelizer (1979, 1985), and later also the influential studies by Marion Fourcade (2009, 2011). Another stream can be identified as French in origin. We are thinking here of the Foucault-inspired studies of governmentality (Foucault [1978] 2007); Gordon, Burchell, and Miller 1991), the studies of the economies of worth linked to Luc Boltanski and Laurent Thévenot (1991), and the major works by Boltanski and Eve Chiapello (Boltanski and Chiapello 2005a; Chiapello 2007). The third stream is the combination of science and technology studies and sociology that emerged from Callon's edited volume *The Laws of the Markets* (1998) and the later *Market Devices* (Callon, Millo, and Muniesa 2007). We can also place valuation studies as part of this stream, be it either in the economic sociology version (Antal, Hutter, and Stark 2015; Beckert and Aspers 2011; Geiger et al. 2014) or in the version that is closer to science and technology studies and actor network theory (see, e.g., Doganova and Kernøe 2015; Dussauge, Helgesson, and Lee 2015; Muniesa 2011, 2014).

One of the ways in which these fields have been moving and evolving is by a bundle of edited volumes that have drawn scholars together around the study of markets and valuations of which especially two volumes have been key to our own work: the sociological volume by Beckert and Aspers's *The Worth of Goods* (2011) and the volume coming more from the side of science and technology studies (Callon, Millo, and Muniesa 2007), on *Market Devices*. One key implication of these works was that the study of economics became less oriented toward critique and historical, ideological, and institutional analysis (for this, see, e.g., Asdal, Brenna, and Moser 2007; Barry and Slater 2002). Instead, this direction of research came to lay stronger claims on having its own expertise with regard to how the economy works. Still, accompanying this move is a way of working and analyzing that is simultaneously in relatively close dialogue with economics. Importantly, however, where the conventional disciplinary field of economics has a strong bend toward formal and abstract, often mathematical models and descriptions, valuation studies and new directions in social studies of markets provide much-needed means for a more concrete and material approach to the study of economics and economy. The discipline of economics is, in a manner of speaking, brought down to earth. One of the ways in which this has happened is via the question of value—to which we will now turn.

THE HEATED PROBLEM OF VALUE

When discussing the notion of value, the work of Karl Marx (1818–1883) is still a standard reference. This is for its own merits, but also because of how it stands in important contrast to the later and today dominant subjectivist theory of value as put forth in neoclassical economics. Whereas Marx ([1867] 2018) saw value as originating in the labor that goes into the production of a commodity, neoclassical economics does not have any such theory of real or objective value that can be traced back to the work that goods are made by. Instead, to neoclassical economics, value is determined by the preferences of the consumers; it is decided in the choices expressed in market actions. The direct link to the production process that was essential to Marx is thereby severed (see Aspers and Beckert 2011; Fourcade 2011).

It is from the angle of neoclassical economics, and not via Marx, that the economic sociologists and market studies scholars Patrik Aspers and Jens Beckert (2011) start, discussing the issue of price setting and from there moving to the more overriding question of value. In fact, it is precisely through the angle of value that they set the stage for a dialogue with and intervention upon economics. Aspers and Beckert (2011) agree with neo-classical economics that the preferences of the consumers determine value. The only problem, they find, is that this approach cannot explain how consumers are attracted to goods or how their preferences are developed. The theory remains silent, as they put it, on the origin of preferences. To answer this question, they find, a broader notion of value needs to be reintroduced. And this is precisely what their approach offers. Instead of staying only with the notion of price, they suggest reintroducing the notion of value that was suspended with neo-classical economics (for the benefit of preferences as they were expressed in market action). To understand markets, Aspers and Beckert (2011, 27) reason, "one must introduce a notion of value that stands apart from price." If not, they argue (2011, 28), there is no way to "judge" prices (see also Vatin 2013).

In Aspers and Beckert's work, two things happen that are of direct relevance to our own endeavor as well as to social studies of markets and valuation studies more broadly. First, the issue of value is reintroduced as an entity to be studied and considered *beyond* the question of value as expressed in the price and, with this, beyond economists' own take on value. The issue of value is placed at the center stage of analysis and is approached in a way that opens it up to sociological analysis. The research problem Aspers and Becket (2011) pose and seek to solve, however, is about understanding markets and consumer behavior—namely, how consumers are attracted to the goods they (possibly) purchase and how they come to their preferences. In their suggestion for how this can be approached, lies the second move of particular interest to us: They turn to the social and institutional structures of markets and what they detect as material devices that direct and act on the preferences of consumers. We can see this as a step away from the abstract models of neo-classical economics, toward a more sociological and device-oriented approach to the understanding of markets. Following Lucien Karpik (2010), one of the examples of such material devices are "judgment devices," that is, devices that assist and direct consumers in doing exactly that, judging with regard to which market actions to take or not. Examples of such judgment devices are critics, guides, and ratings. With reference to Karpik (2000), Aspers and Beckert (2011) point to the quite famous Michelin guide, yet we can immediately start thinking of a myriad of other judgement devices, like the rating systems of smart phone apps like

Uber and Shopee, or the Skrei quality brand that we examine more closely in chapter 6. The orientation toward devices furthermore provokes a view on market acts as entangled with a range of other material artifacts, and to an approach to portraying markets not in the abstract but as material architectures that give shape to, assist, push, and modify how consumers act and what market exchange is about. The latter is exactly what we take to be a rematerializing of economy. Doing this is a key element to our overall endeavor and something the individual chapters of this book expand on and take in many diverse directions.

There are some very clear limits to the rematerialization in the sociological version we have outlined above. This is due to another side of the research program that Aspers and Beckert's (2011) edited volume puts forward. This other side reveals that their concern is not so much about rematerialization, but rather about the construction of meaning and a focus on the *meanings* that goods obtain for actors (Aspers and Beckert 2011, 11). Despite their interest in devices, this is a genuinely *social constructionist* approach. Therefore, it is perhaps not surprising that when choosing which markets to study, they pick markets where, they argue, value is unrelated to and detached from the materiality of the commodity. Three types of markets are mentioned as examples of such materiality-detached markets: financial markets, markets for aesthetic goods, and markets where ethical issues figure prominently. Such markets hold "a special attraction for sociologists," Aspers and Beckert argue, because they are "in very obvious ways socially constructed" (Aspers and Beckert 2011, 30). They furthermore find these markets to be particularly interesting because they serve the objective of developing a general sociological theory of valuation and the pricing of goods (Aspers and Beckert 2011, 31). Their interest is to understand how market preferences are socially constructed and helped and guided by material devices. To accomplish this, they become concerned with that which is distinctly social and in where the social is made to appear most strongly. The question of nature or the natural world is simply not on their research agenda. To the contrary, in this approach value and valuations are interesting to study to the extent that they are *detached* from nature and materiality more broadly. In sum, Aspers and Beckert (2011) do provide an interesting opening toward a rematerialization of the economy, but this materialization is limited to the role material artifacts or devices play in shaping the meanings consumers attach to goods.

There are other works that rematerialize the economy by emphasizing devices in market analysis, though doing this somewhat differently. Even prior to *The Worth of Goods* (Aspers and Beckert 2011), scholars from more of a science and technology studies and actor-network theory background published the edited volume *Market Devices* (Callon, Millo, and Muniesa 2007). Markets become here a more thoroughly material landscape, and the research problem is different from the social constructivist take above. Rather than the social construction of *preferences*, the issue is the material-semiotic construction of *markets*. The concern is less with how devices shape the meanings and preferences of consumers and more with how devices are involved in *rendering something economic*. The market, rather than consumers, is put center-stage.

Much in line with earlier actor-network theory-oriented research, like *Science in Action* (Latour 1987) and *Laboratory Life* (Latour and Woolgar 1979), *Market Devices* is oriented toward the material arrangements by which actions are enabled and happen. The difference is that here, in social studies of markets, the issue is not with the making of scientific facts but with the making of markets. The very notion of market device is a way of referring to the material and discursive assemblages that intervene in the construction of markets (Muniesa, Millo, and Callon 2007). Similar to how science, in the broader field of science and technology studies is approached as a material, technological, *and* social affair (see, e.g., Latour 1990), the economy is understood as material, technological, and social. And just like science and technology studies broke off from technological determinism, the approach to markets here is breaking off from economic determinism or, in other words, a market logic. Last, but not least, in similar ways to how laboratory studies within the actor-network theory approach moved science from being understood and analyzed as a predominantly cognitive and theoretical affair to being approached as a material or material-semiotic issue, the same move is happening here, in the understanding of markets.

In science and technology studies, the notion of constructivism points to the manufactured character of scientific knowledge. Science does not simply mirror the objects and realities studied, this implies, but takes an active part in shaping them. This is not least due to the devices or instruments through which scientific knowledge is, precisely, constructed (see Haraway 1988; Knorr-Cetina 1981; Latour and Woolgar 1979). The turn to the study of markets as framed by Michel Callon, Yuval Millo, and Fabian

Muniesa above should be understood in parallel ways. Economics, we learn here, does not simply describe the economy, but actively takes part in enacting and manufacturing it. In other words, economics is *performative* (see, e.g., Callon 1998; Muniesa 2014; MacKenzie 2006; MacKenzie, Muniesa, and Siu 2007). And herein lies also the link to market devices: it is by being inscribed in such material and discursive devices that economics is made to perform (see Callon 2007): performing markets.

What, then, is a good example of a market device? One of the examples being put forward is the shopping cart—understood as a material device that is turned into a market device and in this capacity reconfigures what shopping is and what shoppers can do (Callon, Millo, and Muniesa 2007; see also Cochoy 2007, 3). After all, can a market exist without a set of market devices, Callon, Millo, and Muniesa ask rhetorically, before listing a series of other examples: analytical techniques, pricing models, purchase settings, merchandising tools, and trading protocols. This is a rematerialization of the economy, and more profoundly so than in the sociological approach we outlined above. The research question is here not that of solving how actors develop their preferences for market goods assisted by and equipped with material devices, but how the market is being realized by way of devices. These devices are combinations of the material *and* semiotic; they are simultaneously discursive *and* material objects (see also Boldyrev and Svetlova 2016; Callon and Muniesa 2005; Muniesa 2014).

In this literature, however, the orientation toward devices is still quite closely linked up with the (human) actors' capacities to act. Together, the devices and the actors become actor-networks, to use an earlier category of actor-network theory. The agenda moreover remains with the devices and does not seriously include the materiality of the commodities that these are set to work on, nor nature, or the natural more broadly. We get to know much more about the specific tools of markets than about, for instance, how biological organisms such as a cod affect economic practice (Asdal 2015). Even when Çalışkan and Callon (2009, 2010) later have extended their approach to include also the things being exchanged, they reason that a prerequisite for things to be made into exchangeable goods, is that these things must be rendered passive (Çalışkan and Callon 2010, 5; see also chapter 6).

Given the initial and strong interest in more-than-human agencies and capabilities in science and technology studies and actor-network theory, this is a little surprising. Even more so, if we take into account that it was Michel Callon who wrote the famous so-called scallop-paper (Callon 1984) where he pleaded scholars to explore and take nonhuman agency into account. The paper on the domestication of scallops in the French bay of St. Brieuc and the scallops' ways of acting on and interfering with these practices and the researchers' efforts, became a standard reference for the actor-network theory approach. Like in this book's story of the Atlantic cod, the scallops were turned from a wild into a farmed species. Also very much in line with what we will show in the case of the cod, the scallops did not readily lend themselves to this enterprise. The scallops refused to cooperate with the scientists' initiatives. Yet, interestingly, whereas the agency of scallops is here brought into the story, the very market for these scallops, or the architecture of markets more generally, is taken as a given. In Callon's scallop-paper the consumers are already waiting for this commodity. The scallop does not have to be modified to please the taste of their potential buyers, and the value of this product is apparently not an issue. Interestingly then, all that is here taken for granted has later been subjected to scrutiny in a program for investigating markets, but in doing this, the issue of nature and nonhuman agency, was left behind. Consequently, we find ourselves in a situation of a missing conversation between two equally important strands of research. On the one hand, there is the body of work that brings out ways in which the nonhuman can act with or without, on, and alongside the human. On the other hand, there is the field of valuation studies and social studies of markets. The latter does wonderful work in bringing out the performative aspects of economics, and the technologies and devices of the economy, but has so far rendered nature as being rather passive. It does little to act on and influence how economy is manufactured and practiced.

What Callon's (1984) seminal work on the scallops animates, his work on markets and market devices renders passive. If we are to understand nature-made economy, we need to re-combine the two and extend the description and analysis to the very things being exchanged, considering how they actively play out in market actions, but also in the numerous operations that are involved at the many different sites before landing there. To renaturalize studies of economy means that we need to take entities like scallops and cod fish and their nature-entanglements into the study of processes of economization and, furthermore, to do so in ways that pay notice to them precisely *because* they are material, at quite important points both lively and living, and always also acting upon the economy. In other words, we turn the research problem, as it was put by Aspers and Beckert (2011) above, on its head.

We take an interest in the cod and its economization precisely because of its materiality and "naturality," and suggest that by doing so, we can learn what market work is, how preferences come into existence, but we can also learn about economies—and by extension also societies, more broadly. By expanding our attention from the social world of consumers, the devices that assist them, and the device-made realities of markets to the rich nature-worlds that markets and commodities are made from, we can learn about preferences, but even more about dependencies, precarities, and the vulnerabilities of nature as well as economy. We can begin to ask, is this a good economy? What ends do this economy serve and what troubles does it cause—or perhaps help solve? To grow this awareness, the objective of the research, we find, must be different from that of seeking general sociological rules. Rather, we must delve into the complexities of nature and economy relations, in their becoming, in their workings, and in their failings, too. By acknowledging these relations as contingent and fragile, we may open economic analysis not simply toward investigating the *social* world of markets but also toward how nature and the living are made to take part in the calculative and speculative endeavors of the economic and, in turn how the economy and nature are *co-modified*.

FROM ECONOMIZATION TO "VERSIONS OF ECONOMIZATION"

A key term in social studies of markets is the notion of *economization*. To embark on the endeavor of bringing nature into the analysis of economy we will let this notion assist us, yet also suggest how to modify and develop it further. The notion of economization is quite intimately related to that of market devices, which we introduced above. The term is inspired by the notion *agencement*, a term Callon, Millo, and Muniesa (2007) borrow from Gilles Deleuze to signpost that there is an intimate and intricate relation between the actor and the thing. It alerts us to that subjectivity cannot be grasped as something external to the device, but is enacted *with* the device: an actor-network that renders things, processes and behaviors economic (Callon, Millo, and Muniesa 2007). Importantly, the emphasis is put on the "rendering" and not on any substantive definition of what "economic" should mean. The economic is rather the outcome of a process of, exactly, economization. This implies that the economy cannot be grasped as a pre-existing reality, nor as regulated by a specific inner logic. In fact, there are no economic laws to discover, nor any specific mechanisms that can be expected to be triggered by specific actions. Instead, we are invited to consider the economy as an ongoing achievement. This ongoing achievement is economization (Çalışkan and Callon 2009, 370; Callon 1998).

The notion of "marketization" was later added to the social studies of markets vocabulary. It is intended to describe the particular form of economization that organizes "the conception, production and circulation of goods" (Çalışkan and Callon 2010, 3). With this, the notion of "economization" is specified as a program for studying markets which understand this process as a step-by-step procedure (cf. Çalışkan and Callon 2009, 5). It starts by "pacifying goods," moves to "marketizing agencies," "market-encounters," and "price-setting," and then ends with "market-design and maintenance." This approach to the study of markets and economies considers the economic from a quite open and empirical angle, rejecting definitions that already at the outset determine what the economic is. Rather than explaining how and why actions happen by pointing to a given or underlying societal context and structure, the interest is with how practices enact, add to, and transform the world (see Callon and Law 1982; Asdal 2012). Economizations are performative; they enact and manufacture the economic. To this it must be added, though, that *how* the world is transformed, in the sense of what it is transformed *into*, is left more open. By reading *Market Devices* we do nevertheless get a little closer to this question. It seems undeniable, the editors reason, that "in so-called advanced liberal societies, 'economic' often refers to the establishing of valuation networks, . . . to pricing and to the construction of circuitries of commerce that render things economically commensurable and exchangeable" (Callon, Millo, and Muniesa 2007, 3). Yet, sometimes, they admit, the economic means other things altogether, for instance, to be saving and to be careful.

The effect of demonstrating how economizing is profoundly a practical procedure, a procedure that happens by material and discursive means, is to open up economy to concrete, empirical scrutiny. The challenge now, we find, is to move onwards, from the relatively thin definition of economic provided by referring to valuation networks, pricing, and circuits of commerce that render things commensurable and exchangeable. And second, to be able to situate economizations and examine concretely and empirically how economization processes can vary and change over time. A third challenge is to grow an awareness of how economizations are enacted into ordered and patterned realities, and how different versions of economizations can act out in tension and in conflict with one another. In this book, we introduce the notion of "versions of economization" as one of our main entry points for pursuing how nature is taken into the economy. Such versions of economizations, we suggest, happen by a range of different little tools and procedures, at very different sites, as part of quite different and changing political and economic programs, and with quite different outcomes for the natures and the economies in question. Importantly, in underlining "versions" regarding economy (Asdal 2014), we are not approaching versions as different *perspectives* on the same reality. Instead, inspired by the work of Annemarie Mol (1999, 2002), versions are to be understood as realities that are differently enacted in various settings, and where different versions can form a relational space, can contradict one another, interfere, or align. Different versions can also be differently formed or stabilized over time (Asdal 2011).

Compared to the sociological version of social studies of markets we first addressed by considering the work of Aspers and Beckert (2011), the economization approach is no longer only about studying markets and seen from the side of consumers and their preferences. Economization is more broadly about how things come to be economic. Another merit to this approach, is to consider the material and discursive composition of the economic, rather than concentrating on the construction of meaning. The economization program is still quite programmatic. The quite detailed, step-by-step procedure we referred to above risks moving along the same trajectory as the discipline of economics itself, toward becoming quite abstract and formalistic and focusing too narrowly on a restricted understanding of what economization is and by which means it happens. If the analysis of *economization* and particularly *marketization* is turned into a programmatic procedure, we risk subduing both too much and too little under the economization umbrella. Too much in the sense of reducing too much to the calculative and exchangeability aspects of the economic. Too little because the economization approach is defined as how market actors act. What this book shows is that a range of different actors and arrangements is involved, not the least state agencies, The means and arrangements by which such agencies act and work are equally also quite diverse. The programmatic procedure furthermore signals a quite linear understanding of economization. The economization procedure comes across as an always forward-moving process wherein entities are inevitably increasingly economized. Due to this, the economization approach is less well rigged to capture how nature is taken into the economic in a myriad of different ways, is not passively economized, and may also act back, interfere with, and trouble the economic. It is toward this end—that of making space, analytically and empirically, for this—that we let the notion of co-modification assist us in this book's endeavor to investigate nature–economy relations. It is also toward this end that we add *valuation* to the notion of economization and replace the notion of market devices with *tools of valuation*.

TOOLS OF VALUATION, VALUE ORDERINGS, AND VALUATION ARRANGEMENTS

To equip our analysis with the capacity to capture how nature is made part of and is involved in versions of economization, we need to further expand the analytical vocabulary from that of *market devices*. Our analytical approach, we find, need to be able to capture that the relevant tools not only are attached to the market but to a range of other sites, too, such as labs, production sites, parliaments, ministries, research councils, and other public agencies. For this, we need a term that is site-neutral and that does not already at the outset of the analysis discriminate between different sites or define the tools involved in economizations as belonging to a specific (market) activity. Importantly, the tools that we examine do more than to construct markets, indeed, also when they *are* involved in market work or broader processes of economization, the tools we encounter do a range of things: they assess, judge, consider, praise, price, care for, acknowledge, recognize, count, enlarge, lift, tone down, bind, make space for, and more. Taken together, we suggest, what they do is *value*. They are involved not simply in economization, but in valuations, broadly conceived, and the tools involved work as *tools of valuation*. This is, then, one of the key notions that this book puts forward and works with, which in turn means that the study of economization is extended toward that of valuations. This move is important, as sometimes valuations challenge economization.

Other times, valuations take part in composing nature-economy entanglements and modify and alter versions of economization. Taken together, this means that, similar to how tools of valuation encompass more than the tools of the market, valuations range broader than economization.

In turning to valuations, we acknowledge that nature gains significance and worth by a range of practices other than simply pricing- and calculative procedures. Just as prevalent are rich and complex descriptions, ecological vocabularies, how consumers and other actors asses and regard nature and nature issues, and how nature is made to appear, by being mapped, drawn, placed on agendas, or made part of laws, rules, restrictions, and, most notably, document procedures tied in with political and bureaucratic bodies and agencies. Such valuation procedures, however, also impact market work, innovation strategies, and processes of economization. Omitting this from the analysis, treating economizations as if the calculative were simply the logic of it, we reduce too much to a limited version of what economy is and thereby also fail to understand how actions and operations of economization, including market making, operate. We also fail to understand how economizations are restricted, limited, opposed, and questioned. As we show throughout this book, nature does not encounter economy only through being economized, and nature is frequently taken into account by more-than-market means. Sometimes economizations provoke or render visible a nature that works for other ends, that is, than to become economized and turned into economy.

Taken together, economization does not come in only one format. Nor is economization a linear procedure. It can unfold by a complex set of tools and comes with tensions and conflicts. Moreover, economization does not simply work *by* tools of valuation; these tools of valuation work *on* the entities that are rendered economic. These entities, in turn, come with their own affordances and agencies that act on economizations as well as on their ordered outcomes. Different versions of economization compose entanglements of economy and nature differently. They produce different formations of society and are not neutrally involved in economization but spur the question and engagement in what good economizations and economies can be (Asdal et al. 2021). Also in this latter respect, they are tools of valuation.

The fourth and fifth notions in our methodological vocabulary, adding to the notions of versions of economization, co-modification, and tools of valuation are *value orderings* and *valuation arrangements*. In making this move toward valuations we build from Çalışkan and Callon's (2009) approach to economization, but extend, modify and also empirically ground it. In putting forth the notion of value ordering, we furthermore build on John Law's classic study *Organizing Modernity* (1994; see also Law et al. 2014) and its concept "modes of ordering." Rather than there being "order," Law reasons, there is ongoing ordering work, an ordering work that nevertheless becomes patterned by "modes" that shape them. The argument Law here makes, further resonate with the work of Annemarie Mol (2002) and the tension she identifies with Foucault's concept of the episteme. There is no longer any belief in Foucault's episteme, Mol reasons in *The Body Multiple* (2002). Foucault's way of delineating societal orders that thoroughly shape realities and subjectivities alike, is understood to be too strong and with too few cracks, openings, and fragilities to them. In contrast, "modes of ordering" as well as "versions" are constructivist terms that seek to capture how realities are patterned but always with an "on-the-move" character. Different value orderings, as we denote them, may overlap and encounter one another in tension as well as in alignment. It is toward the pragmatist underpinnings to the study of value and valuations we will now return, and in doing so, further situate our own study in relation to this literature.

PRAGMATIST UNDERPINNINGS TO VALUATION

We remember from above how Aspers and Beckert (2011) argued, against economics, that to understand markets one must introduce a notion of value that stands apart from price. We fully agree with what must be done: we must reintroduce to the study of markets and, more broadly, to the study of economy a broader notion of value. Narrowing this down to the study of meaning, as in Aspers and Beckert's program, is however an unnecessary limitation, as is also that of staying with the question of how consumers develop their preferences. Still, this is not to say that we wish to return to the Marxist theory of value. To go completely to the "other side" and sidestep the market side to value for the benefit of production and labor is not a satisfactory strategy. Several of the chapters in this book show that enhancing the market value of cod commodities, or helping it along by a series of tools of valuation, is clearly a part of the valuation-picture. Our objective is anyhow not to build a general theory of value. Our objective is to be able to detect, describe, and consider how valuations are done, the tools and apparatuses by which this happens, and so how valuations come to matter and shape how economies are ordered. This is not to exclude meaning or "attraction," but rather to expand it—toward the nature objects that are being worked on, and also toward the normative and political aspects to how economies are arranged and come into existence—hence also the notion of "the good economy" (Asdal et al. 2021), a notion that speaks to how economies often not only seek to order the production and trade of goods, but also come are involved with versions of the good and normative ends.

The American philosopher and pragmatist John Dewey (1939) and his theory of valuation has inspired and underpins many of the contributions to the field that has become social studies of markets (see, e.g., Antal, Hutter, and Stark, 2011b, 2015; Geiger et al. 2014; Stark 2011b; Trébuchet-Breitwiller 2015). More recently it has also engendered a whole new domain of research under the heading of, precisely, valuation studies (Dussauge, Helgesson, and Lee 2015; Muniesa 2011; for an overview, see Asdal, Doganova, and Fochler, forthcoming). Our turn to valuations as a methodological approach and to the notions of tools of valuation, valuation arrangements, and value orderings are likewise inspired by the pragmatist approach in the tradition of Dewey. Rather than approaching value as something subjective, as something we "hold," or as the objective character of things, we follow Dewey in considering value as practical and something to be grasped in its unfolding—as *valuation*. The question of value is then redefined as a practice, simultaneously making it an object to be investigated empirically. Underlining the shared etymological root of notions like price and praise, valuations are rendered an interdisciplinary endeavor, across, for instance, economics and sociology. Via such pragmatist moves, the economy and the economic are re-opened to empirical examination. Our ambition is to move such pragmatic approaches to valuation into the study of economy in its nature relations, and to make the study of the economy and economics consider—just as seriously as human preferences—the environment, the affordances of nature, and the once-living turned into things being exchanged. There is of course already literature that has embarked on related endeavors, and in valuation studies, too (see, e.g., Çalışkan 2010; Doganova and Kernøe 2015; Friese 2015; Heuts and Mol 2013). Yet doing this not only raises new questions or calls for new methodological takes and approaches. It also, we suggest, invites new conversations.

BRINGING NATURE'S FRAGILITY CENTER STAGE

There are many places, studies, and notions from where the conversations pushed forward by our pragmatic approach to nature valuations can be started. One of the many works that have inspired us is the edited volume *How Nature Works: Rethinking Labor on a Troubled Planet* by Sarah Besky and Alex Blanchette (2019). With their notion of "troubled ecologies" they draw the reader into the fragility of nature or, more concretely, how this fragility becomes apparent in the labor that we both live by and demand from nature. This is not an abstract or lofty issue or discussion, but materially and concretely present in laboratories, factory farms, plantations, thinning rainforests, militarized borders (Besky and Blanchette 2019, 6), and, we can add, the net pens with masses of farmed cod that this book in part is about. The trouble that Besky and Blanchette refer to is an important reminder of whose question we are seeking to answer when bringing out and becoming aware that these places are far from stable systems of accumulation, but are instead characterized by fragile relations. Their question, as ours, is not how, at what cost, or most efficiently we can make nature work toward our ends, but rather how to trouble such sites of labor extraction and, in our case, demonstrate how economies are very concretely made from them. With this, the book joins forces with a nice body of studies from sometimes very different genres, but which share a critical curiosity toward investigating nature commodities. Some of these are particularly tuned in to following how goods of nature are turned into goods at markets. John Soluri's (2002) classic piece on bananas and their co-modification (to draw on our own notion) with the judgments and habits of American consumers is one of these, as is Roger Horowitz's (2004) study "Making the Chicken of Tomorrow."

The chicken is especially intriguing in relation to aquaculture. For whereas the cod was envisioned to be modeled on the salmon, the salmon was initially modeled on the chicken. And, of course, there are studies of fish, too, that are caught and taken to markets, like Alexander Dobeson's (2016) study *Hooked on Markets*, Marianne Lien's (2015) *Becoming Salmon* on the domestication of salmon (see also Law and Lien 2013) and of fish that are becoming what they are at the farm, and Petter Holm and Kåre Nolde Nielsen's (2007) fascinating piece on fish that are turned into economic objects by being caught in the regulatory tool called "individual transferable quotas." In fact, Holm and Nolde Nielsen's study was published as part of the volume *Market Devices* (Callon, Millo, and Muniesa 2007) that we have referred extensively to above, and so fish have already entered the field of social studies of markets. There are other examples within this field of active nature-objects, too. Most notable is Koray Çalışkan's (2010) book *Market Threads: How Cotton Farmers and Traders Create a Global Commodity*. With the main objective of understanding how global markets work, this understanding being developed through the lens of cotton markets, but also the fields of cotton, Çalışkan goes a long way not only to rematerialize the economy, but also to renaturalize it. If not as lively as the cod fish, cotton also can be quite troublesome, does not lend itself easily to market exchange, and offers an interesting case on one of the issues with which this book is also concerned, namely, the *pricing* of commodities. The very growing of cotton, its "production," Çalışkan shows, entails a simultaneous engagement of relations that consist not only of exchange and production but also of a series of activities that make up a rich and undertheorized world of encounters and struggles among pests, children, merchants, immigrant workers, women, *khoulis*, farmers, cotton, cows, *gamusas*, economists, ginners, *elchis*, and others (Çalışkan 2010, 20). Çalışkan's study quite literally brings economy down to earth. Interestingly, the cotton fields in his study are never so pacified as the "goods" and "commodities" in the second part of the article by Çalışkan and Callon (2010), where they underline quite strongly that objects need to be made passive to become exchangeable market goods (see also chapter 6, where we discuss this further).

We should be careful not to insist that our objects of research that once were swimming, living, growing, or reproducing are kept, methodologically, alive throughout our studies. In following the production of soybeans, Kregg Hetherington (2019, 2020) carefully reminds us of the fact that farming always involves killing (Hetherington 2019, 52). With a point of departure in the production of soy the question is not *if* it kills, Hetherington writes, but what the relations of living and killing are. An *intensification* of killing might just as well be the relevant issue (Hetherington 2019, 42). The present book further underlines Hetherington's point, and the ensuing chapters will show that even if we insist on the agency and "activeness" of the cod fish throughout, there is killing and slaughtering, as well as trouble and procedures for how to deal with the dead at almost all the sites that this book explores. Caring and killing (cf. Law 2010a) do not go together by all means and prompt questions of what a good economy is, was, or can still become.

FROM A DENATURALIZED TO A RENATURALIZED ECONOMY?

In calling for a renaturalization of the economy, there is one key work that must be addressed: Margaret Schabas's (2005) book *The Natural Origins of Economics*. Until the late Enlightenment, Schabas writes, the natural and the economic realms were one and the same. It was not before the first half of the nineteenth century that the concept of the economy emerged as an autonomous entity, economy gradually being identified as a distinct entity subject to operations of human laws and agency. Schabas describes this process as a denaturalization of the economic order. Before this, money, trade, and wealth were all understood to be a property of the physical world. For instance, when the physiocrats—a French school of economic thinking that saw agriculture and land as the origin of wealth—formulated how they understood the question of circulation and reproduction of wealth, they did so in terms of a natural order, where nature was understood to be the prime mover of economic processes. Schabas contrasts this situation with the contemporary, which she describes as being the radically opposite. "Economists today," she writes, "study a world that is essentially detached from the processes of physical nature" (2005, 12). Instead, this is a world that is made to stand apart from the physical world, as if it were made by humans: "Matter no longer constrains or determines, as it did for the Physiocrats and the classical economists. No longer does land scarcity or the stationary state loom on the horizon. We find a conception of the economic order that is more or less severed from physical constraints. Wealth, or utility, is granted an unprecedented ability to expand" (2005, 16). To be sure, when we, in dialogue with Schabas's history of the denaturalization of economics, employ the notion of *re*naturalization, the argument is not that we can, or even wish, to return to a place where the social and the natural order are the same. As we are throughout this book concerned with the question of agency, also agencies in the sense of political or hybrid institutions or arrangements, the quest is not to have the economy to simply dutifully and unquestionably reside in a natural order taken as given and decided on by a few (cf. Latour 1998, 2004). Also, our point is not to argue that nature is a pure and clean category, unmediated by action, concepts, and human prints and hands (Asdal 2008, 2003). As will become clear, for instance in chapter 4, we rather show how particular versions of nature are manufactured in encounters with economy. Our quest is about making these re-belong, to bring economy and nature together, and quite literally to bring economy down to earth. From there, we can start attending to how these relations are to be knit and done and to how shared troubles can be handled.

Schabas (2005) laments how economic attributes have lost their physical dimensions. Wealth, she argues, became a nonmaterial entity in the mid- or late nineteenth century, and capital is now essentially "a claim on the future and is thus defined in terms of the discounting of time" (2005, 13). It is from this notion of capital that Liliana Doganova (2018, 2023) has demonstrated how the technology of discounting was developed in a quite down-to-earth way. Discounting, Doganova shows, was developed in the nineteenth century in close connection to forest management and the question of when to cut the forest to secure maximum return in the long term. With this, the notion is linked to sustainable development. In this sense, Schabas's claims may be too strong. There *are* links, even if modest ones, and they can be traced also at the discursive level. In fact, one of the things this book shows is how concepts pertaining to nature and concepts pertaining to economy are still overlapping and meet in complex and sometimes surprising ways. To have an eye for this, however, demands that we watch and tread carefully.

In chapter 5, we touch upon the work of David Harvey (2010), who has portrayed capital as a bloodstream. Besky and Blanchette (2019, 10) draw on a more recent contribution, of Jason Moore (2015) and his portrayal of capitalism as "not imposed on an ecology," but "itself an ecology, one that animates relationships between machines, grains, and chickens or with shaping these creatures in the flesh." Our take is to study capital as an empirical term, following how it moves, not describing it so much as that which animates others or other things, but as that which is itself being animated, moved around, qualified, praised, and made to take on new capabilities, strengths, and behaviors. Consequently, we treat capital first as a practice, and we approach this thing called capital quite concretely and empirically, and then observe how it is made into something quite vivid. The vividness or animateness of capital, however, is not due to some inherent quality it holds, be it blood-like or ecology-like, but is situated and emerges by way of its relations to other entities and practices. We furthermore approach capital not only in the meaning of funds, or money to invest, but also in the meaning of *bio*capital, and we do this in conversation with a series of works that have investigated "the bio" from this angle of capital.

Importantly, the notion of biocapital introduces a distinction between capitalizing on the matters of the living and other forms of "capitalization" (Muniesa et. al 2017), involving other, differently constituted materialities (see chapter 3). An important strand of this research on biocapital is one that is often cordoned off as "feminist," showing how biological entities, by virtue of being *re*productive, introduce a new dynamic to production (Cooper 2008; Franklin and Lock 2003; Franklin and Ragoné 1998; Thompson 2000, 2005; see also Murphy 2017). This, in turn, has spurred questions over what counts as labor, with concerns being raised especially over how women and their bodies are part of asymmetrical relations of product-making and profit-seeking (Cooper and Waldby 2014; Herzig and Subramaniam 2017), but also, more broadly, over how to understand the myriad ways in which nature works (Besky and Blanchette 2019).

We can agree with the critique put forward by Besky and Blanchette (2019, 13) that terms such as "biocapital" (Rajan 2006), "promissory capital" (Fortun 2008; Thompson 2005), and "lively capital" (Haraway 2008; Rajan 2012) "call attention less to 'how nature works' under capitalism than to how the generative (reproductive) capacity of living beings (human and otherwise) is both determined by capitalist relations and captured by regimes of accumulation." Yet, the challenge also goes the other way around too, toward opening up for how nature *and* capital work, investigated not as predefined logics of accumulation but as empirical matters and movements. Kaushik Sunder Rajan's (2006, 2012) conceptualization of biocapital does, for instance, quite eminently demonstrate how a concept such as capital, when examined empirically and situated in different cultures of science, industry, and government, cannot be pinned down to the one logic or definition, but must be understood in terms of plurality. In Rajan's work, biocapital is consequently rather eclectically defined. The various approaches to biocapital must also be seen in the context of which research problems one aims to solve, and the sites and materials one is working with and from. For instance, we do address in this book how capital is made to circulate (see especially chapters 3 and 5), but our endeavor is directed more toward how biological entities are made to take on capital qualities in the first place. This means that it is the very act of modifying and transforming something *into* biocapital that interests us, and we argue for examining this as concrete and empirical practices.

To accomplish becoming biocapital, we emphasize in this book, "the bio" needs to be reared, grown, and nurtured— it must adjust, conform to, and "agree with" requests from the market and demands to keep investments high, but costs low. This is not to say that species comply with such demands. The notions of rearing, growing, and nurturing imply agency and part-taking also on the side of the species, and are a call to attend to how this agency also works on and affects how "the bio" can be made economic. Yet again, this is a co-modification endeavor. Becoming biocapital is about co-modifications—practices where both "the bio" and the capital are modified and worked on. Examining such practices empirically and in efforts to manufacture biocapital defy arguments of capitalist logics of accumulation that work by their own force, subduing the bio into neat, pacified and complying objects. As this book shows, efforts to modify the bio into a form of capital, to *biocapitalize*, may fail, and such efforts come with costs that are not only about the loss of money but also about environmental problems, illness, ethical dilemmas, and losses, sometimes quite massive, of the organisms that are subjected to these biocapitalization exercises.

In this, we would like to link up with Anna Tsing (2015), who follows a very different species than ours, namely the matsutake mushroom, "to explore indeterminacy and the conditions of precarity"—or, as she also puts it, "life without the promise of stability" (2015, 2). It is only with an appreciation of precarity understood as an earth-wide condition, she underlines, that we can observe the situation our world is in (2015, 4). In this way, Tsing brings life to the center stage of her analysis, but life and forms of living that come without the promise of stability. The matsutake mushroom Tsing follows has taken her, as the title of her book signals, to the end of the world, and to look for life in capitalist ruins—ruins that are, as she puts it, spaces abandoned as asset fields (2015, 6–7). The Atlantic cod has similarly allowed and indeed requested us to travel widely, across oceans and net pens in Norwegian fjords, into state bodies, and to Chinese mega markets. Yet, different from the ruins of Tsing, we are taken here into the enormous efforts made to produce value from the living, the massive investments that are put into a species, the programs of innovation it is subjected to, and that work to make the Atlantic cod behave like biocapital. By following these programs and the huge potentialities that the cod is heralded and praised for, we are simultaneously exposed to the precarities, instabilities, and vulnerabilities of such endeavors. We are also exposed to that which is not in ruins: political valuation procedures, arrangements for valuing cod movements, valuation arrangements that work toward "more-than-market prices," and concerns that are raised as collective and shared. There are investments in good economies, too, and not only investments, but debates and conflicts regarding what the good economy is and how goods can best be pursued.

Much like the human–animal studies that our book learns from and adds to, the sites our book examines—the ocean and the economies and values that are made from it—are not academically unexplored. Taking on the great transformation of the ocean, we join a still small, but already rich and now rapidly growing literature that seeks to counteract a long-standing neglect of the ocean in the social sciences and humanities—a "terracentrism," in the wording of Eric Paul Roorda (2020, 1). A turn described as both "oceanic" (Deloughrey 2016, 32) and "blue" (Braverman and Johnson 2020, 2) has spurred scholarship on the ocean as a socially constructed and historical space (Demuth 2019; Laloë 2016; Steinberg 2001; Wigen 2006), on the ways in which the ocean is being explored scientifically (Adler 2019; Helmreich 2009; Schwach 2013) and exploited economically (Arbo et al. 2018; Eikeset et al. 2018), on the ontologies of the ocean (Hastrup and Hastrup 2016), its interspecies relationships in view of climate change (Deloughrey 2019) and on legal regulation (Braverman and Johnson 2020). This also includes publications that, like ours, focus on the one species, such as Jennifer E. Telesca's *Red Gold: The Managed Extinction of the Giant Bluefin Tuna* (2020), the hugely popular *Cod: A Biography of a Fish That Changed the World* (1997) by Mark Kurlansky, or *Floating Coast: An Environmental History of the Bering Strait* (2019), wherein Bathsheba Demuth meticulously attends to how the whale has helped shape environmental and human history.

GLOSSARY

Throughout this chapter, we have underlined the conversations our book seeks to engage in; the key theoretical, conceptual, and empirical contributions it is indebted to; and the contributions that have assisted us in delineating our own concepts and methodological approaches. We have shown where our questions and problems depart from earlier work in social studies of markets, and why and how our concepts, analytical approaches, and methodological moves take on different routes. Both in the introduction and above we have explained our paths toward the concepts that we will be employing throughout the ensuing chapters of this book. In this final section of the chapter, we briefly summarize this methodological topography, with entries ordered conceptually.

LITTLE TOOLS: This notion refers to material-semiotic entities, technologies, or artifacts that in and of themselves are modest, small, and act locally, but that by being part of larger machineries and apparatuses, by their movement, and by their combination with other such tools, perform crucial work.

TOOLS OF VALUATION: This notion builds further on the notion of little tools. They are material-semiotic entities, technologies, or artifacts that in and of themselves are modest, small, and act locally, but that by being part of larger machineries and apparatuses, by their movement, and by their combination with other such tools perform valuations. Tools of valuation may operate across different sites, be they market sites, state bodies, or hybrid agencies. They are tools that perform valuations, which can be done in a range of different ways. They can work on markets, but they can also be linked to politics and political procedures. Tools of valuation can work by calculations, such as numbers, accounting, growth estimates, and prices, as well as through the qualitative, the normative, the narrative, and other modes. They may appreciate, downplay, uphold, praise, prize, or acknowledge. They may be material-semiotic tools like maps, different document species, brands, or surveys. Precisely how and what is being valued by various and distinct tools is open for empirical analysis. Tools of valuation may be part of economizations and of producing versions of economization, but this is not necessarily or always the case.

VALUATION ARRANGEMENT: This notion is about how various valuations, including a variety of or a given set of tools of valuations, are organized and/or bundled together in more encompassing formats than in a tool of valuation alone. This can be through relatively formal material arrangements; through hybrid formats that are set up across, for instance, market actors and state bodies; through arrangements inside already established organizational forms; or through arrangements made for limited periods of time and for working on specific tasks. For instance, a market site can, with its distinct architecture, pricing procedures, and other tools of valuation, be considered as a particular valuation arrangement, as can the sites of production and their procedures for ordering and enhancing the value of goods. Valuation arrangements can also be formal organizations, like the Norwegian Seafood Council, which we analyze in this book (see chapters 6 and 7).

VALUE ORDERINGS: This notion denotes the stabilization of a valuation, and points to an order that is the outcome and consequence of valuations. Value orderings are stabilized valuations that are enacted through a series of actions or procedures, but they are not necessarily stable over time. They need to be worked on and reenacted to stay stabilized. Different and competing value orderings may enter in tension and conflict with one another, and established and new value orderings can clash. Value orderings can be predominantly economic, such as the exclusive economic zones that we describe in chapter 2, and which can be considered a value ordering that is also a version of economization and tied to tools of valuation and valuation arrangements, such as the system of licenses for petroleum exploration and extraction. Value orderings do not need, however, to be economic.

VERSIONS OF ECONOMIZATION: This notion builds on, extends, and modifies the notion *economization.* Versions of economization is a way of describing and analyzing how economizations play out in different versions. That economizations are enacted in different versions means that to economize may vary with regard to time, site, scale, and means. That there are versions of economization points to that economizations are materialized in different outcomes that take the form of patterned realities that are open for change, but also in part are stabilized. Versions of economization do not happen by material arrangements alone but can come into being through complex interplays between the objects and entities sought capitalized upon, market actors, and the state. A version of economization can be about creating a market, but it can also be about a Parliament's way of ordering an industry, or about reworking innovation schemes. The three notions above—tools of valuation, valuation arrangements, and value orderings—are different elements that can make up distinct versions of economization. Versions of economization also happen by way of co-modifications.

CO-MODIFICATION: This notion builds from the notion of commodification, the turning of something into a commodity and rendering something economic. The notion simultaneously makes a twist to the concept of commodification and is used to alert us to the practices and work involved in *modifying* entities: markets, consumers, and capital, on the one hand, and things that are economized or capitalized, on the other. In this book, we focus on co-modifications in nature-economy relations, but co-modifications can also span more widely. Further, such modifications meet in *co*-modifications where things and markets are made to work on one another. It thereby describes a double entendre, and it is directed toward describing and analyzing how turning something into market objects or, more broadly, rendering something economic is a relational and not a pregiven, linear, or uniquely social process. It is rather something that we can examine more openly. The notion is meant to work broadly and to include, in this book, the co-modifications of nature and economy in its multiple shapes and formats, but could in the context of other studies involve other instances where the initially non-economic and economic meet. The notion is also meant to encompass not only markets and market work, but also the co-modifications of production and capital. Co-modifications can happen by various tools of valuation, through different valuation arrangements, and take different shapes in relation to varying versions of economization.

BIOCAPITALIZATION: This notion points to the practices involved in valuation processes aimed at turning something into new forms of *bio*capital. It points to how nature and the living are never straightforwardly already biocapital, but must be turned into biocapital, and how this can involve a series of different tools, practices and operations. Biocapitalization is consequently a version of co-modification in the sense of it being conditioned on modifying both "the bio" and capital. The notion is useful for investigating empirically how this happens at different sites, by different tools of valuation, and by way of various valuation arrangements, and how this can involve different procedures, such as rearing, growing, and nurturing biocapital. The notion *biocapitalization* is also useful for addressing how these are open-ended practices, that may fail.

GOOD ECONOMY: This notion is an analytical tool for investigating how the economic and versions of economization are entangled in versions of the good. Consequently, the notion speaks not only to the practical workings and outcomes of specific versions of economization but also to what it seeks to achieve and enact as being "good economy." The notion is intended to be useful for analyses of how the normative, the qualitative, the moral, the political, and different versions of value and that which is considered of worth are part of economization. It is meant as an analytical tool to examine how different versions of the good can be integral to versions of economization; may be the outcome of versions of economization, valuation arrangements, and value orderings; or may be in tension and conflict with distinct versions of economization. The notion addresses how economization is not a neutral activity, but that economizing also prompts examinations of worth that go beyond the straightforward economic. The notion does not imply that economies *are* good but is intended to work as an analytical tool for investigating how versions of the good are or aim to be part of economy and economizations.

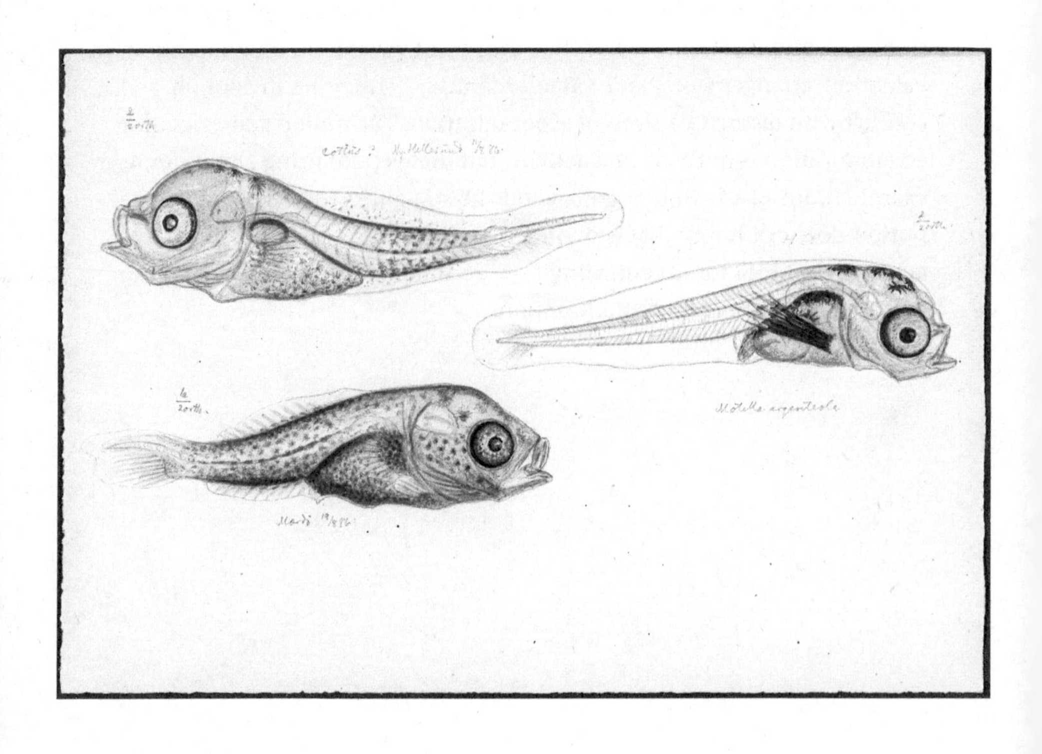
Cottus ?
Motella argenteola
Mardö 19/8 86

2 RADICAL REVALUATION: ECONOMIZATION AT ALL DEPTHS AND SCALES

The front cover image of the OECD foresight report, *The Ocean Economy in 2030*, tells it all: the ocean and its beings are about to be exploited at all depths and scales (OECD 2016): fishing, farming, drilling, mining, researching, prospecting, and surveillance—on the seabed, in the water column, on the surface, along the coast, and far out to sea. The opportunities are seemingly endless. Petroleum activities, for instance, are imagined as being moved to "ultra-deep water and exceptionally harsh environments" (OECD 2016, 3), while bio- and nanotechnologies are presented as creating new opportunities for exploiting and harvesting the biological resources of the ocean. Under the purview of "blue growth" the oceans of the world are being set up to undergo a transformation that will forever change them. Once a place of sailors and fishers, the oceans of the world are set to be crowded by new and expansive industries.

As discussed in the introduction, the OECD is only one of many agents pushing for ocean economy growth. Other intergovernmental bodies such as the EU and national governments—including that of Norway, but also the United States, Canada, China, and other countries—have recently presented their own strategies for how to profit from the so-called blue economy (Reinertsen and Asdal 2019; Voyer et al. 2018). These differ in many respects but have in common that they frame the ocean as a space of untapped economic opportunity, a space where many and new industries can thrive. Indeed, the visions being presented portray the ocean as a place where its users—old and new, human and nonhuman—coexist without much tension or conflict. When cast within the prose of forecast reports

FIGURE 2.1
Front cover of the Organization for Economic Cooperation and Development (OECD) foresight report, *The Ocean Economy in 2030* (OECD 2016).

and innovation strategies, conflict becomes something else—a challenge to overcome, a situation to learn from, perhaps even an entry point to creating future profit—all while ensuring that economic growth answers to the demands of environmental sustainability and reduced climate emissions. Not only the economy, it seems, but also the environment and climate will benefit from this growth. This is, as it puts itself forward, a problem-free, win-win economy.[1]

ECONOMIZATION—OTHER VERSIONS BY OTHER MEANS

The win-win vision could easily be contradicted by pointing to the already existing growing pains of the ocean economy, such as biodiversity loss or the numerous conflicts that surround extractive industries. What we explore in this chapter, however, is a different type of counter-story: the so-called "oil–fish conflict" that arose between the fisheries and the petroleum industry in the late 1960s, when offshore drilling was introduced in the North Sea. Rendering economic the previously unknown pockets of oil and gas resting below the seabed, offshore drilling is in many ways what set in motion the great economization of the ocean. It involved a radical revaluation and reordering of the ocean as a space of economic opportunity and growth, and it is one of the very first examples of the type of ocean crowding put forward by the OECD, the EU, and other organizations. In this chapter, we explore the tensions that arose when the fisheries and petroleum industry began to compete over the same ocean areas, and we look at the tools used in efforts to resolve their conflict. Contrary to the OECD's way of valuing the ocean by use of economics and by ascribing *monetary* value to its so-called "natural assets" and "ecosystem provisions" (Asdal et al. 2021; Nebdal 2019), we find that the valuations performed in opening the ocean as space of multiple, but also conflicting industries rested on the tools and orderings of the state. The valuations available to us when assessing different versions of economization, the chapter thereby shows, are not limited to the expert judgements of economics. They can just as well be handled by other means—by way of political procedure, policy documents, and public inquiries that work as tools of valuation and perform value orderings both in opposition to, in tension with, and in exchange with the calculative practices of economics. As much as propositions such as that put forward by the OECD are about economization, this tells us, they also perform orderings, in this case, value orderings that renders the ocean economic in new and extended ways, by new means.

In this chapter, we start out by considering how, in the 1960s and 1970s, the oceans of the world underwent a radical reordering. Much of the legal basis for today's ocean economy was established and from belonging to a relatively open and shared international space, large parts of the ocean were drawn into national territories and legislation. In Norway, the outcome was two spatial orders. Geographically these overlapped, but they contained distinct and, at times, competing versions of economization. First, a new border for the ocean's *seabed* was established in 1965. Huge parts of what had previously been international seabed, including its underlying resources, were included in the sovereign territory of Norway. Subsequently, it was reordered by way of a geographic grid structure fit for licensing rights to petroleum exploration and extraction (see figure 2.2). Second, in 1977, Norway successfully claimed a new border for the so-called water column, extending it from 12 nautical miles from shore to 200 nautical miles (see figure 2.3). Unlike the seabed border this did not entail an expansion of the nation's territory. Instead, the water column border was designated as a so-called "exclusive economic zone" to be controlled by the state. The state could now regulate the outtake of living marine resources from this zone, such as the valuable cod fish, thereby creating a new and differently constituted ocean commons. Spatially and materially, this signified a radical revaluing and reordering of the ocean, opening it for economization in distinctly new ways.

From concentrating on how these new and differently economized ocean spaces come about, the chapter turns to examine how their use quickly came into tension with one another, creating the oil–fish conflict. The fishers feared oil spills and environmental damage, but most prominently, they claimed compensation for the loss of valuable fishing grounds to the petroleum industry. Through a close reading of policy documents commissioned by the state, and which drew on consultations with and representation by the fishers, the chapter shows how the material artifacts of maps, in intimate entanglements with the procedural capacities of policy documents, became tools for valuing and ordering the ocean as a space of two economies: an expanding petroleum industry and the already existing fisheries. Grounded in the tools and procedures of the political, these valuations and orderings make apparent that economization is not a one-way process; as new entities of nature are brought into the economy, economic practice is also made to adjust. Rather than seeing the economy as a prefigured, preformatted system, this suggests, we must consider how it is actively modified and reordered, and how this is done by a range of tools and entities, the tools and procedures of the political being one of them, the workings of nature another. In paying attention to this, we address in this chapter the role of spatial configurations in valuation struggles and examine how a series of maps take part in the economization of the ocean. We also show how these maps can act as a form of counter-valuing to economization procedures, making present and thereby also valuable already existing economies of the ocean.

The oil–fish conflict speaks in interesting ways to Marion Fourcade's (2011) seminal study "Cents and Sensibilities: Economic Valuation and the Nature of 'Nature.'" Studying the legal settlement of compensation for damages caused by two major oil spill incidents—one being caused by the supertanker *Amoco Cadiz* off the coast of Brittany, France, in 1978, the other by the supertanker *Exxon Valdez* in Alaska in 1989—Fourcade untangles the intricate relationship between how the value of nature is perceived and the tools of economics applied in expressing this value in monetary terms. The oil–fish conflict was not primarily about oil spills, but like these cases, it involved claims for compensation, made by North Sea fishers who experienced the rapidly expanding petroleum industry as inflicting damage on and impeding their business. In the cases studied by Fourcade, the issue of compensating for damage to or loss of nature was settled in the legal system, and by use of tools of valuation belonging to economics. The procedure used to settle the oil–fish conflict, we will show, was quite different: The government, in consultation with the fishers, opted to keep the matter of compensation *outside* the legal system and *inside* the procedures of politics. Ultimately, the value of lost fishing grounds was considered by quite different tools of valuation, like mapping procedures and documents tied in with the valuation arrangements of politics and administration. Rather than valuing the fisheries, the fish, or fishing grounds in monetary terms and thereby settling the issue by providing the fishers with economic compensation, one created the so-called Oil–Fish Fund. This scheme was mainly directed at "improving" the efficiency and profitability of the fisheries and has a certain likeness to the innovation schemes of today (see chapter 5). Interestingly, and as we return to in chapter 3, it also funded efforts toward an altogether different version of economization, namely that of breeding and rearing a new and domesticated cod.

RADICAL REVALUATION I: NATIONALIZING THE SEABED AND ITS SUBSURFACE RICHES

The map shown in figure 2.2 is one of the most influential maps ever drawn for Norwegian territories. It is, in many respects, an icon of the ocean's great economization. Published in April 1965, this map—which in everyday parlance is referred to as "the shelf map"—is the first of what today counts as thirty-five such maps. Updated regularly, the shelf maps show where and by whom petroleum exploration and extraction is taking place on the Norwegian continental shelf. At its peak, in 2001, Norway was the world's third largest exporter of oil and largest exporter of gas (Bjerke 2009), but at the time of the 1965 map's publication, the prospect of striking oil was still only theoretical. Several international petroleum companies had shown interest in exploring for oil, which prompted the government to develop a legal and regulatory framework. Passed in Parliament in April 1965, the Exploration and Extraction of Subsea Petroleum Deposits Act details how both exploration and extraction should take place. Under paragraph 11, it is stated that licenses to explore for and extract petroleum can be acquired for a so-called block (Innst. 1965, 5), an area that in the 1965 shelf map is represented by gridded lines cutting across the ocean.

At the time, Norwegian lawmakers had no experience to draw upon when developing legislation for an offshore petroleum industry (Innst. 1965, 4). The committee appointed to draft the law therefore chose to study the legislation of other countries, among these the United Kingdom, where a block system had already been put into place (Innst. 1965, 30).[2] The actual size of the blocks—500 square kilometers—was determined by consulting the petroleum companies, and with the purpose of excluding "non-serious" license seekers (Innst. 1965, 32). The blocks were moreover cast upon an already existing cartographic representation of the ocean, as they were placed within so-called quadrants, squares whose length and breadth equal one degree longitude and one degree latitude. Each quadrant contains twelve blocks, and the size of the blocks is also given with reference to degrees of latitude and longitude.

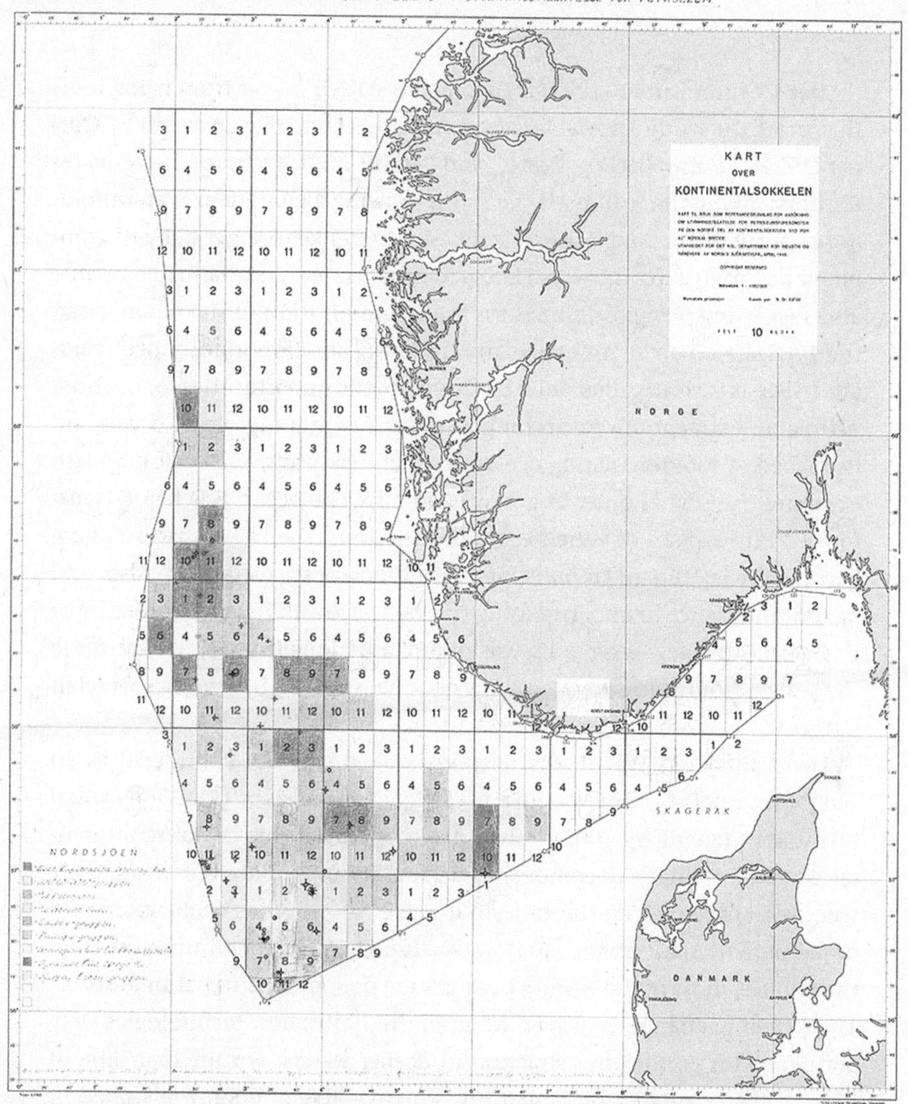

FIGURE 2.2
Map of the continental shelf, produced in 1965 and published by the Ministry of Industry and Craft. The squares in the map are so-called quadrants and designate an area consisting of twelve blocks for which a license for petroleum exploration and extraction can be acquired. In this version of the map, the quadrants have been hand-colored, the colors signifying the company that holds rights to the given quadrant. The outer lines of the map represent the borders between Norway, the United Kingdom, and Denmark (map retrieved from Norwegian Petroleum Directorate, https://norskpetroleum.no).

Maps, Bruno Latour (1992) argues, achieve their power from being tools that make the world knowable, navigable, and claimable. Or as put by Corner (1999; see also Kitchin, Dodge, and Perkins 2011, 6), mapping activates territory, giving the map itself the status of being a conduit and an unfolding of possibilities. And it was exactly, we would like to suggest, by its combined way of making the world knowable, navigable, and claimable—while also suggesting new possibilities for the economization of the ocean—that the 1965 shelf map became such an interesting and important tool of valuation. For as is clearly illustrated by this map, the opportunities for offshore petroleum extraction were accompanied by a reordering of the ocean and its seabed. A tool for valuing ocean space in new ways, the shelf map also expresses the spatial order of a new economy. The ocean was being transformed into a place of vested economic interests, the block spaces having become the legitimate grounds for a new industry to operate on. Also, and of key importance to understanding this radical revaluing and reordering of the ocean, this new geography was part of a nationalization, transforming the seabed from being an open, international space to becoming sovereign territory.

Up until World War II, the deep-sea seabed was not considered as an economic resource, nor was the ocean considered as a place for capital investment (Steinberg 2001). It had played a crucial role in transcontinental empires, military operations, and long-distance trade but was largely seen as a space outside the terrestrial places of progress, civilization, and development. Apart from coastal waters, usually no farther from shore than three miles, most of the world's oceans were one, huge international space. This began to change, however, when in the 1940s, new technologies were developed for petroleum extraction in deeper waters. For the first time in history, the possibility of capital investments "fixed" upon the seabed of the deep sea was introduced—in the form of pipes, wells, and platforms—and with this came an opportunity for extending industrialization from the land to the ocean. The petroleum industry thereby introduced to the ocean what the nineteenth-century political economist David Ricardo (1817) defined as fixed capital: the real or physical assets that go into production—say, a property, a building, or the machinery of a factory. Because of this development, the three-nautical-mile limit that most states accepted as the outer border of their ocean territories was soon to be politically challenged. This, in turn, had great consequences not only for how the Norwegian map is drawn, but also for how the state would go about when ordering the value of its ocean territories.

The first such challenge was made by then US president Harry Truman, who in 1945 issued two proclamations expanding US authority beyond territorial waters (Steinberg 2001). This was soon followed by declarations by other countries eager to secure potential petroleum reserves, many of which were recently decolonized countries. In 1958, the first United Nations Conference on the Law of the Sea was held in Geneva. The conference resulted in four conventions—the 1958 Geneva Conventions on the Law of the Sea—stating the right of coastal states to exploit the resources of their continental shelf—that is, the subsea extension of their land territories up to a depth of 200 meters (Treves 2008). By 1963, the Norwegian government had declared sovereignty over the Norwegian continental shelf, setting the new border not at a specified place, but "as far as the depth of the ocean allows for exploitation and exploration of the natural deposits, regardless of the otherwise applicable ocean boundaries, nevertheless not beyond the median line in relation to other states."[3] Subsequently, the state began to develop regulation for exploring and exploiting subsea natural resources. Two years later, though not without conflict, Norway had come to an agreement with the United Kingdom and Denmark on where to divide the North Sea continental shelf between the respective countries. This division is also represented on the 1965 shelf map, in the form of two of the three border lines that delineate the grid structure of the blocks. For whereas the uppermost line follows the latitude of 62° north, and with that the northern limit set by the Norwegian Parliament for the area available for petroleum exploration, the other two lines are borders. Drawn toward the west and the south, they signify where the newly extended, sovereign territory of the Norwegian continental shelf stops.

THE NOT SO EMPTY BLANKS ON THE MAP

As mentioned above, the 1965 shelf map not only was a tool for knowing, guiding, and making navigable; it was also a means of delineating and organizing economic possibility. Shortly after the passing of the law in Parliament, the first round for obtaining a block license was announced by the Ministry of Industry and Craft.[4] Altogether, 278 blocks were made available, which makes this the most comprehensive licensing round to have ever been held for the Norwegian continental shelf.[5] This is also reflected in the map depicted in figure 2.2, where some of the blocks appear to have been colored by hand—each of the different license-holding companies having been given its own color. As the handwritten title inscribed on the top of the map states, this is not just a map depicting the block structure of the North Sea. It is a "Map That Shows the Awarding of License to Extract Petroleum,"[6] and with this how the ocean was now being taken over by new actors—actors with a license to drill. The shelf map affirms a new value ordering of the seabed, reordering it as a gridded space of state-controlled blocks licensed out to commercial actors, while simultaneously acting as a tool of valuation. Revaluing the seabed as a place over which individual company rights can be asserted—some blocks already taken, others not—the shelf map holds the promise of chance, speculation, and profit for those willing to invest and explore.

This economization and, simultaneously, nationalization of the seabed speaks quite interestingly to both Karl Polanyi (1944) and Karl Marx's ([1867] 2018) political economic analyses of England in the eighteenth and nineteenth centuries. Writing about the transition to what he identifies as a market economy, Polanyi devotes an entire chapter of *The Great Transformation* to the discussion of "land" and how it was reordered by industrialization, transforming countryside into cities and cities into class-divided landscapes of capital and work. This was a transformation, we could add, that in many cases involved the parceling of urban land into grids not dissimilar to those of the 1965 shelf map (Rose-Redwood and Bigon 2018). Marx, for his part, describes the enclosure of the commons. The privatization of what had previously been the common lands of peasant communities, Marx observes, transformed these lands into mere commercial commodities.[7] The reordering of the ocean's seabed into license blocks is in many ways similar. An open and free space is effectively closed. Still, it is also quite different from the transformations Polanyi and Marx describe. For there had not been any previous economic or subsistence use of the seabed, *and* the blocks remained public property: the seabed is not purchased by the companies but licensed to them by the state. Block licenses are awarded on the basis of applications from interested companies and not, for instance, to the highest-bidding company. Not a market, but the *state* decides which companies are allowed to develop which areas.[8] This is, then, a version of economization where not only a new industry emerges, but also a new sovereign territory, the two bound together in a revalued, reordered ocean space. The ocean's seabed—now considered a state space to control and build its tax base from—becomes a form of capital, a space for securing, strategically, wealth for the nation.

As it appears on the map, the reordering of the ocean into a gridded structure of blocks is quite neat and tidy. Such cartographic tidiness, however, conceals what was a far more troubled ocean reality. For as the petroleum companies moved into the color-coded blocks of the North Sea, a different but overlapping geography was also under negotiation—namely, that of the water column. The seabed and its subsurface riches had indeed been nationalized, but the waters that flowed above—the ocean itself, some would say—had not. And the main users of these waters, the fishers, were far from content with sharing the ocean with the new and fast-expanding petroleum industry. Indeed, from the perspective of already established users of the ocean, the shelf map can also be seen as a tool that values by way of the absences it creates. Neither fishers nor the elevated grounds of the seabed, with the fish banks so attractive to them, are present on this map. The new map of the ocean is drawn, quite plainly, without consideration of these interests and geographies. Instead, what the fishers experienced were new and disruptive activities, displacing them from many of their most valued fishing grounds, a displacement that was soon to be accompanied by the introduction of yet another reordering of the ocean—the so-called exclusive economic zone. This zone, which we now turn to, represents a different version of economization. Geographically, it overlaps with the shelf map, but it works by other value orderings and other tools of valuation.

RADICAL REVALUATION II: ZONING THE OCEAN

The nationalization of the seabed entailed a radical revaluation and reordering of a seabed that so far had not been exploited. The waters above it, however, were very much so. Closest to shore, about four nautical miles out from the so-called baseline,[9] was what we can designate as the national coastal commons and, farther out, an international deep-sea ocean space. The most important economic resource of these waters was fish, and so fishing was the dominant industry. This is not to say that these were entirely open, or uncontested, fishery commons. Rather, and as the history of

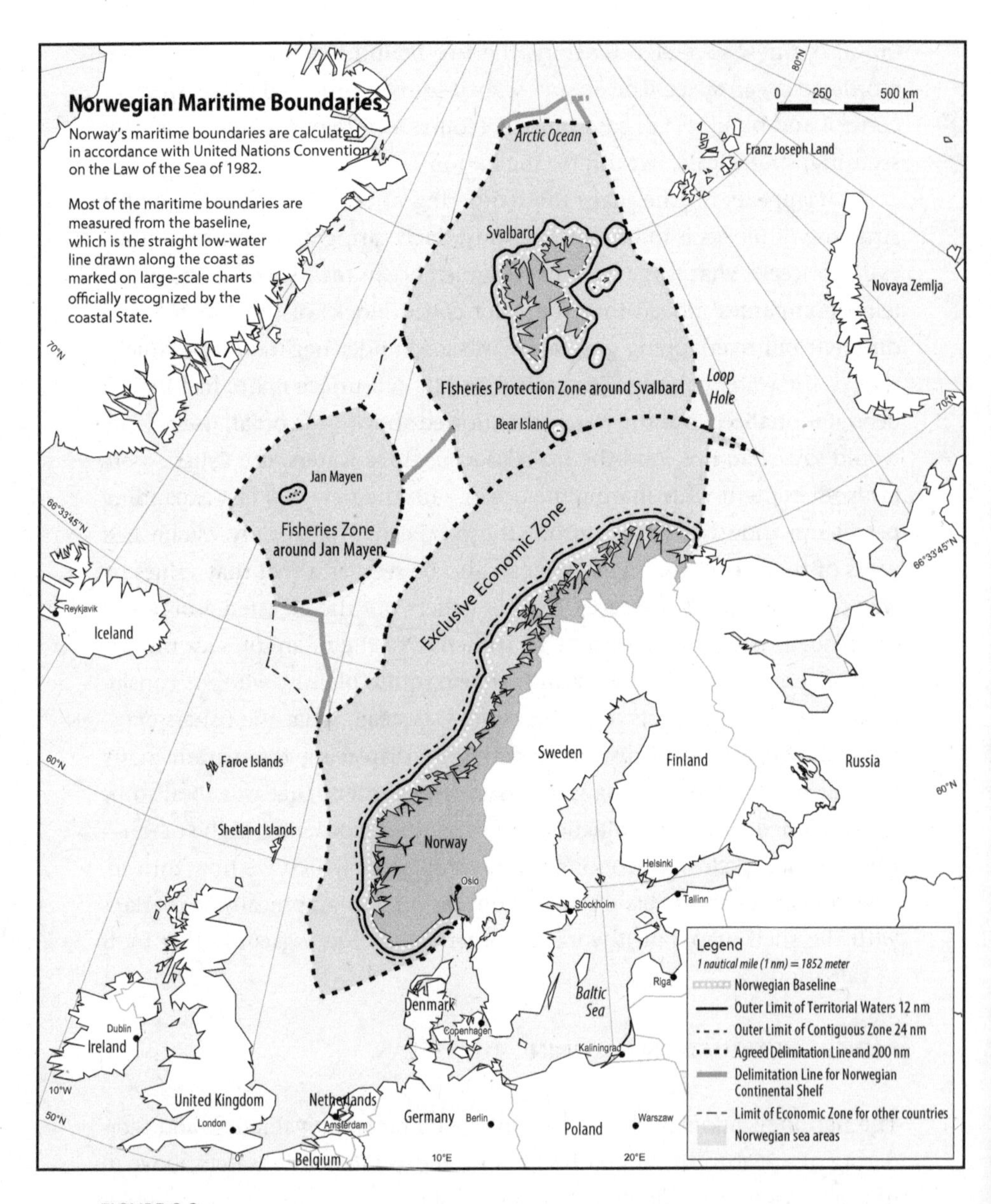

FIGURE 2.3
Norwegian maritime boundaries (Norwegian Mapping Authorities, https://kartverket.no).

especially the coastal commons shows, these have not existed in the "pure" form imagined by Garett Hardin (1968) in the essay *The Tragedy of the Commons*, where the commons is exploited by all and without any regulation. Instead, the commons has been regulated by law as well as by practice and prescriptive rights and has throughout the centuries often been contested. For instance, in the Middle Ages it was usual practice that people along the Norwegian coast claimed specific rights within an ocean area, taking the area as theirs and demonstrating this through use and by actively excluding others from the area (NOU 1986, 44). Properties with a shoreline were seen as continuing into the shallow ends of the ocean, while the areas farther out were a so-called *allmenning*, a commons open to "all men" belonging to the surrounding areas. Parallel to this, restrictions were made on using the ocean as a way of passage—such as the so-called Bergen rule from the end of the thirteenth century, which forbade foreigners to travel the seaways north of Bergen, a prohibition that was kept until the sixteenth century. Records also show that in the late eighteenth century, there were places along the coast where private individuals owned fishing grounds, among others the cod-rich areas outside the Lofoten peninsula and the Varanger Fjord. Along the coast of Sunnmøre, island communities had divided the ocean into so-called *teiger*, strips of ocean belonging to different communities. There were also regional laws for selected fisheries, such as the three laws on the Lofoten cod fisheries passed in Parliament in 1816, 1857, and 1897 (Christensen 2014b, 128–129). In the late nineteenth and early twentieth centuries, however, much work was conducted to coordinate such regional regulations into one national law. The rights to the coastal commons would eventually designate an *allemannsrett*—that is, a "right of all men" of the nation to access and harvest from it, irrespective of property rights or any other entitlements (NOU 1978, 39–40).

These examples in no way cover the whole political history of the Norwegian ocean commons, but do show that, throughout the centuries, it has been organized in different ways. Such previous uses of the ocean have also been key to international negotiations and tribunals over where to draw the border for a nation's exclusive right to exploit marine resources (Hersoug 2005). Such negotiations have been ongoing since at least the nineteenth century, in various formats and fora. Conflict has often run high, which in the case of Iceland and the United Kingdom led to the so-called Cod Wars, which were fought in the late 1950s and continued into the 1970s, as Iceland wanted to close its surrounding waters to foreign trawlers (Store Norske Leksikon, n.d.a). Also other coastal states, including Norway, were in the 1970s expressing great discontent with the presence of foreign fishers (Christensen 2014a). For while the number of fishing vessels was growing rapidly, the fish stocks were dwindling. Years of technological advances in the fisheries had taken their toll on many of the most important commercial fisheries. In Norway, the herring fisheries had collapsed completely in 1970, and from 1972 onward the cod fisheries were also showing signs of a significantly weakened stock (Christensen 2014b).

The concern about overfishing motivated international negotiations. Led by the United Nations Conference on the Law of the Sea, these ran from 1973 to 1982. Before their conclusion, however, nations began to declare that the border of what was to be named the "exclusive economic zone" was to be set at 200 nautical miles, or about 370 kilometers from the baseline of their shores (UN 1982; Christensen 2014a, 65).[10] Iceland declared its 200-nautical-mile zone in July 1976, the United States in April 1976, followed by France and Mexico; and, in October 1976, the Norwegian Prime Minister declared that from January 1, 1977, Norway would also establish a 200-nautical-mile exclusive economic zone (Christensen 2014a, 53–54), thereby extending it from the twelve-nautical-mile border declared in 1961 (Ruud and Ulfstein 2011, 151). With this new extension, the Norwegian exclusive economic zone consisted of 875,000 square kilometers, an area that was four times larger than the mainland.[11] The smaller, coastal commons was as such significantly enlarged, becoming the newly constituted ocean commons. As stated clearly by the Marine Recourses Act, "The wild-living marine resources belong to the community of Norway."[12]

TOWARD A CLOSING OF THE FISHERY COMMONS

The implementation of the exclusive economic zone gives nations the exclusive right to manage and exploit marine resources. This includes the resources in and on the seabed *and* in the waters above (BarentsWatch 2018). By including the waters of the ocean, this version of economization put into play a rather different value ordering than that of the shelf map, as the zoning of the ocean revalues and reorders an already economized ocean commons. The border between the national and international is pushed—from close to shore to far out at sea—constituting the new and enlarged national ocean commons as a resource space controlled by the nation. This nationalization was, however, not absolute. Through a fine-meshed net of agreements some of the foreign fishers retained their rights to fish in Norwegian waters, as did some Norwegian vessels that had previously operated in foreign waters. Also, there were several areas farther north that would remain contested for a long time—with regard to both the seabed and the water column—such as the oceans surrounding Svalbard, Jan Mayen, and the so-called Grey Area between Norway and Russia.[13] However, and as is the case also for other species that move across the borders of nations, bilateral agreements have here been put in place on how to monitor, manage, and exploit the Barents Sea fish stocks. Since 1976, the cod has been the subject of yearly negotiations in the so-called Joint Russian-Norwegian Fisheries Commission, whereby mutual agreement on, among other things, fishing quotas is reached (Joint Fish, n.d.).

Like the gridded structure of the 1965 shelf map, the exclusive economic zone can be considered both as a value ordering and as a tool of valuation. By way of this, the waters within the exclusive economic zone become a form of economic depository, a bank of ocean resources available to and, importantly, under the care of the nation. Inserting a new ocean border, exclusive economic zone cut across old ocean geographies and radically reordered them. Unlike the shelf map, however, this value ordering was not about making the non-economic economic; it shows how an already economized ocean space is reordered and made differently economic. From different modes of ordering emerge different versions of economization.

For the fishers, the implementation of the exclusive economic zone meant that the ocean commons would gradually become more closed (Hersoug 2005). The open, international ocean space had shrunk significantly, which meant that fishing in the rich waters of, for instance, Iceland, Kalaallit Nuunat, and the Soviet Union was no longer available to all. Meanwhile, in Norwegian waters, fishers were soon faced with increased regulation, largely in the form of quotas limiting which species, which amounts, and where the individual fishing vessel could fish. The reasoning behind and the success of this has been fervently contested (Holm 1996), but it is quite clear that amid the petroleum industry's crowding of the ocean, the fisheries were experiencing a closing of the commons on several fronts. Also, and as we turn to now, the fisheries were increasingly competing with the petroleum industry over many of the most attractive fishing grounds. What, then, could the state do, when the economies of an increasingly crowded ocean came into conflict?

TO COMPENSATE FOR THE LOSS OF INTANGIBLE GOODS

By the time that the exclusive economic zone was put into effect, altogether 168 wells had been drilled in the North Sea and three oil fields had been developed.[14] Catches in the fisheries were down by as much as a third of previous levels. The fishers argued this was due to the entrance of the petroleum industry and demanded to be compensated for what they formulated as the "loss of fishing time" and the "loss of fishing grounds"—compensation, in other words, for both time and space. But how to measure the loss of such entities? What, exactly, were "fishing time" and "fishing grounds" worth? How could these intangible goods be valued? And to what extent could one say that the fishers had rights to them, rights that when they were weakened or taken away should—or even could—be compensated for?

In her analysis of the Amoco Cadiz and Exxon Valdez oil spills, Fourcade (2011) examines the compensation settlements that followed in their wake. Both cases were settled within the US court system, but the compensation was far from equal. The size of the oil spill and the scale of damages caused by the *Amoco Cadiz* disaster was considerably larger than that of *Exxon Valdez*, but whereas the *Exxon Valdez* incident cost the Exxon Corporation close to $2 billion, the *Amoco Cadiz* incident cost Amoco about $200 million. In Fourcade's analysis this difference is pinned down to mainly two things. First, the French and American publics valued nature differently. Whereas the French were reluctant to put a monetary price on nature, the Americans were not. They had, in a manner of speaking, different nature cultures and different views on monetizing nature's value. Second, and related to these different positions, different tools of economics were used to calculate the monetary value of the nature damaged by the oil spills. Whereas in the French case one focused on economic losses suffered because of the oil spill, such as reduced fishing or income from tourism, the American case went on to putting a price on the existence of pristine wilderness. This happened by a procedure known in economics as contingent valuation, by which surveys are used to establish a price for goods that are not traded in markets.

When settling compensation for the loss of an intangible good, Fourcade (2011) demonstrates, the tools used to establish and judge its value are of key importance. She further points to the legal system as a key site for studying the economic valuation of intangible goods, precisely because it is very often here that the value of such goods is finally determined. The legal system is, in a manner of speaking, particularly well positioned to make judgment, as the law is equipped with explicit tools for reasoning and justification toward such an end. The oil–fish conflict, however, was taken to Parliament and made subject to judgments reliant not on economics but on established political procedure. Instead of deciding on the proper economic method for calculating damages, the Norwegian government turned to a particular "species" of policy documents—the so-called NOU report. Like legal work, we therefore suggest, the document work of the political is a quite interesting site for studying the contested valuation of intangible goods, the ocean commons being one of them.

WINNERS AND LOSERS IN THE OIL–FISH CONFLICT

Producing an NOU report—the abbreviation stands for Norsk offentlig utredning and translates as "Official Norwegian Report"—is a standard way of approaching complex issues in the Scandinavian political system. This is a "document species" that we discuss more closely in chapter 4—suffice to say for now is that a NOU is, as a rule, the first in a series of "document movements" (Asdal and Reinertsen 2022) often leading up to Parliament voting or deliberating over a final recommendation. It works by assembling assessments, by drawing together actors and their issues, considerations, and evaluations, and by representation as well as by calculation. What follows from a NOU report is often a white paper or a proposition to Parliament, and, finally, a resolution to be acted upon by the government. For the oil–fish conflict, the group appointed to produce the NOU consisted of bureaucrats from the Ministry of Industry, the Ministry of Justice, and the Ministry of Consumption and Administration, but also representatives from the Norwegian Fishermen's Association and the Norwegian Seafarers' Union. The composition of the group tells us that it was not a government-only, expert committee that was given the mandate of assessing "how the oil–fish issue could be settled" (NOU 1978, 6). Instead, the NOU can be considered as a site wherein actors representing different forms of knowledge and interests order and transform complex political issues.

In contrast to the crowded ocean vision promoted by the OECD (2016), of multiple ocean industries coexisting in a win-win ocean economy, the NOU takes as its starting point that there are both winners and losers. The fisheries are quite clearly on the losing side:

> The fishers in the North Sea had the fishing banks to themselves until oil exploration began in 1963. After the first hole was drilled in the German sector, drilling operations followed in the British sector in 1964 and in the Norwegian sector in 1966. At the turn of the year 1977/78 altogether 1,192 wildcat wells had been drilled in the North Sea, of which 748 were in the British sector, 254 in the Dutch, Danish and German sectors and 190 in the Norwegian part of the North Sea. . . . Altogether 104 fixed installations have been placed in the North Sea, and altogether 2,690 kilometers of pipelines have been placed on the seabed of the different countries' sectors. . . . The oil activity has led to a new situation in the North Sea. Whilst the fishers used to be almost supreme on the ocean, there is today extensive oil activity in the form of drilling and development of findings. The relationship between the new and old users of the ocean has caused problems, and the fishers have generally appeared as the losing party. Large areas where there had previously been fisheries are today closed for the fishers because of the placement of drilling and production platforms, with their associated security zones, because of pipelines and because of the littering of certain ocean areas. (NOU 1978, 5)

In extensive detail, the NOU report assesses this win-lose situation by describing the already existing relations between the petroleum industry and the fisheries. It establishes that the fisheries are of great significance, not only to the Norwegian economy but also to maintaining the many fishery-dependent settlements along the coast. To settle the oil–fish conflict, the NOU reasons, is therefore of value to society at large. Yet, another imperative also emerges here: the government plans to extend petroleum exploration northward, beyond the line drawn on the 1965 shelf map at latitude 62° north and into the Arctic waters of the Norwegian Sea and the Barents Sea. Then, as now, these oceans were home to the most important commercial fisheries of Norway, and the conflicts of the North Sea did not bode well for a frictionless expansion. Nor did it help the government's plans that in 1977, only months after the establishment of the exclusive economic zone, a major oil spill occurred in the North Sea. Environmental disaster was avoided, as the oil spill never reached shore; it remained in the North Sea, where it dispersed in a matter of weeks. Despite this not too detrimental outcome, the oil spill incident made it clear to the Norwegian government that before a northward petroleum expansion could happen, one had to learn from and settle the conflict between the petroleum industry and the North Sea fishers (NOU 1978, 26). At stake was the North Sea fishers' compensation, but also the relative powers of the fisheries and the petroleum industry. The issue of being present and occupying what to the one industry was a drilling site, and to the other fishing grounds, was thereby entangled in the question of compensation.

In the NOU report, the fishers' losses stand out as largely spatial: pipelines, which the petroleum companies had failed to bury, ran along 2,960 kilometers of the seabed, "and not a few of these kilometers run over rich fish banks" (NOU 1978, 32). Platform locations tended to "coincide with the richest fishing banks" (NOU 1978, 32), displacing the fishers from their most favored place of business—the elevated grounds, or banks of the ocean where nutrition is easier to come by and therefore more fish assemble (Hommedal 2021). And finally, there was severe littering of the seabed, "partly by littering around boreholes and partly by littering along the supply boat routes" (NOU 1978, 6). The litter could damage the fishers' equipment, and the fishers had therefore begun to register it in their sea maps, so that it could be avoided in the future (NOU 1978, 7). The litter, however, did not always lay still and could drift to a new location. It was therefore difficult to know which litter was what, in that the same object could be registered as being in many locations (NOU 1978, 29). On the fishers' modified maps, the litter had, virtually, multiplied; as an obstacle to be avoided, the one piece of litter could in fact become many.

Amid the pipelines, platforms, and litter assembled by the NOU the ocean emerges a space within which there are *places*—places where the formation of the seabed makes them particularly valuable to the fish and the fishers, but also, and increasingly, places that are occupied and industrialized, the new petroleum industry fixing itself onto the seabed. There is no place in the NOU where the petroleum industry's encroachments are problematized as being an environmental problem. The problem is exclusively enacted as being about a downturn in earnings and an increase in costs for the fishers. In other words, the problem is economic. It is not nature, but the fisheries that have been damaged. And insofar as nature is valued, it is being valued for its entanglements in a nature economy. The ocean is consequently not enacted as an object that has value in itself—a form of existence value that can be represented by pricing it—but as something that *yields* value. This is quite different from the *Exxon Valdez* case analyzed by Fourcade (2011), where one put a price on, exactly, the *nature* damaged by the oil spill, but not so dissimilar to the *Amoco Cadiz* case, where economic losses were focused on. Different from both these cases, the NOU did not move to calculate the monetary value of the fishers' loss, be it of fishing time or fishing grounds. Instead, what it produced was a series of maps that located the value of the fisheries in specific ocean places and, further, articulated the two ocean economies as being in a spatial conflict.

COUNTER-VALUING BY MAPS AND NUMBERS

As part of assembling the NOU, the NOU committee had asked the regional fishery managers of the Ministry of Fisheries to assess the fisheries of their respective regions in "a business economic and regional context" (NOU 1978, 21). Their answer came in the form of maps and numbers. The assessment made by the fishery manager of the Troms region includes eight maps altogether (figures 2.4–2.11). All these concentrate on the rich fishing area Tromsøflaket, which lies in the northernmost part of the Norwegian Sea. The first map stands out, however, as this does not show fisheries, but what a shelf map for this area would look like (figure 2.4). The quadrants numbered 7117–7120, the title of the map tells us, are already "areas opened for further examination" (NOU 1978, 18). The ordering of the ocean by way of blocks fit for licensing is thereby extended—on paper and in practice—from south to north.

On the other maps provided by the Troms fishery manager, the area "opened for further examination" is still drawn in but is now accompanied by depictions of where fisheries take place and of the volumes of their catch (figures 2.5–2.7). They show where there is line fishing and drift net fishing for cod, drift net fishing for saithe, and fishing fields in use by trawlers (figures 2.8–2.11). Cutting across the shaded lines of the rectangle that demarcates the petroleum exploration area, the lines demarcating the fishing areas are drawn not in the fashion of cartographic principles, but in the manner that fish swim. Curved and winding, the lines of these maps show exactly what is at stake: fishing grounds—or to be more specific, *good* fishing grounds. In contrast to what appears on the gridded shelf map as empty blocks for the taking these maps enact an ocean filled with already existing economic activity; an ocean being valued as already economized.

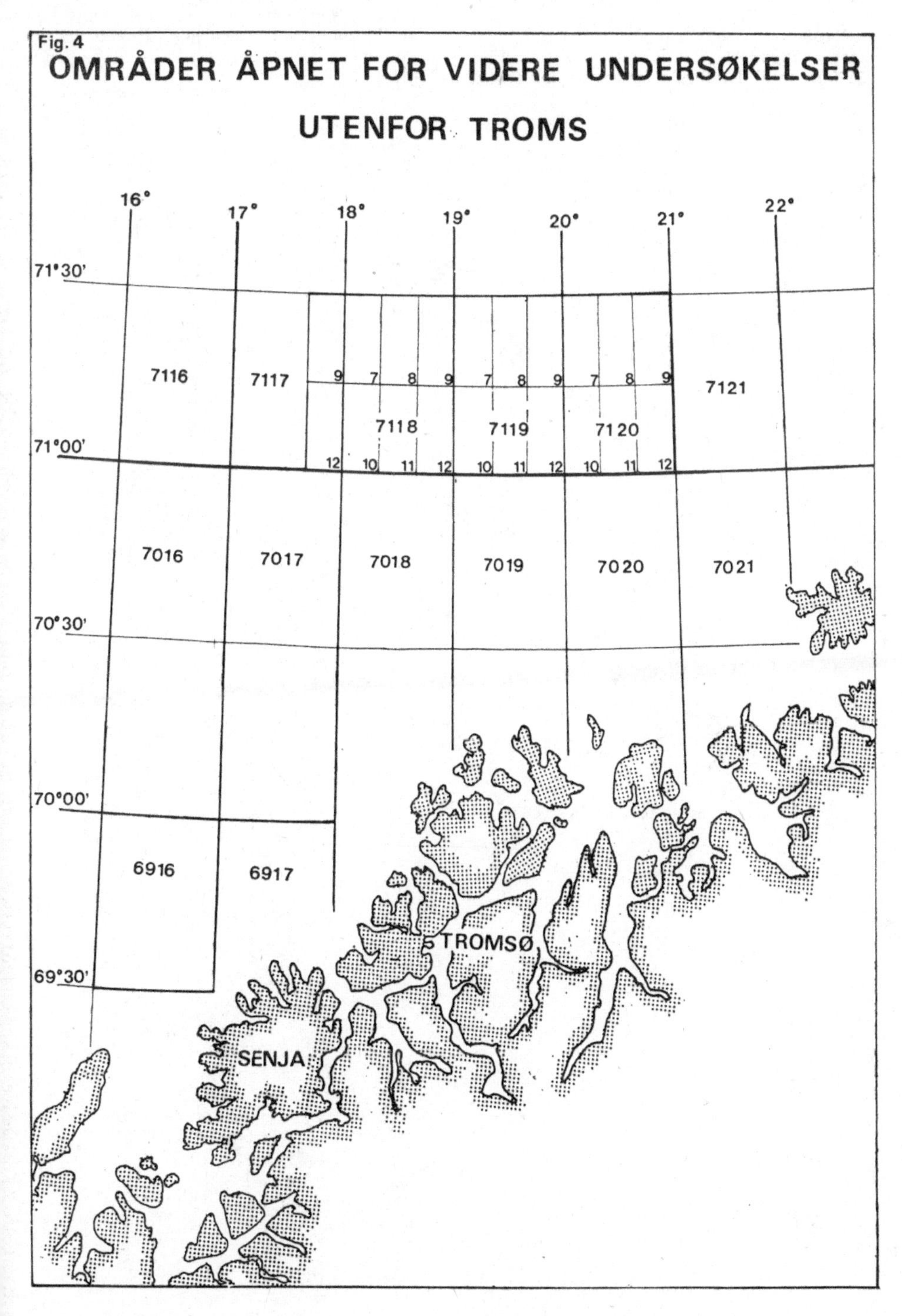

FIGURE 2.4

Excerpt of the shelf map, depicting "Areas opened for further examination outside Troms" (NOU 1978, 18). These are the block areas of quadrants 7117–7120.

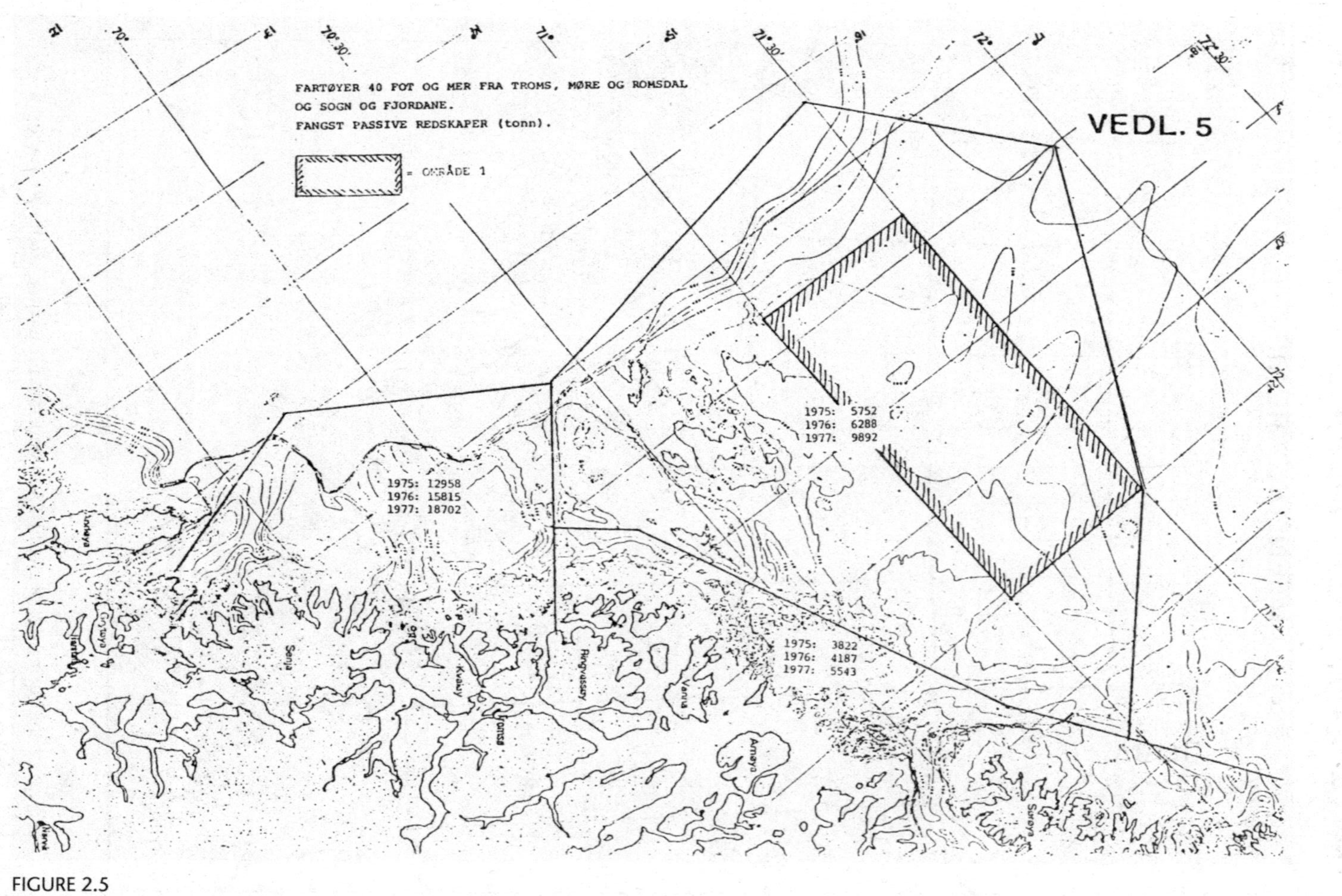

FIGURE 2.5
Catches in tons, by use of passive fishing gear, in vessels over forty feet from Troms, Møre og Romsdal, and Sogn og Fjordane (NOU 1978, 64).

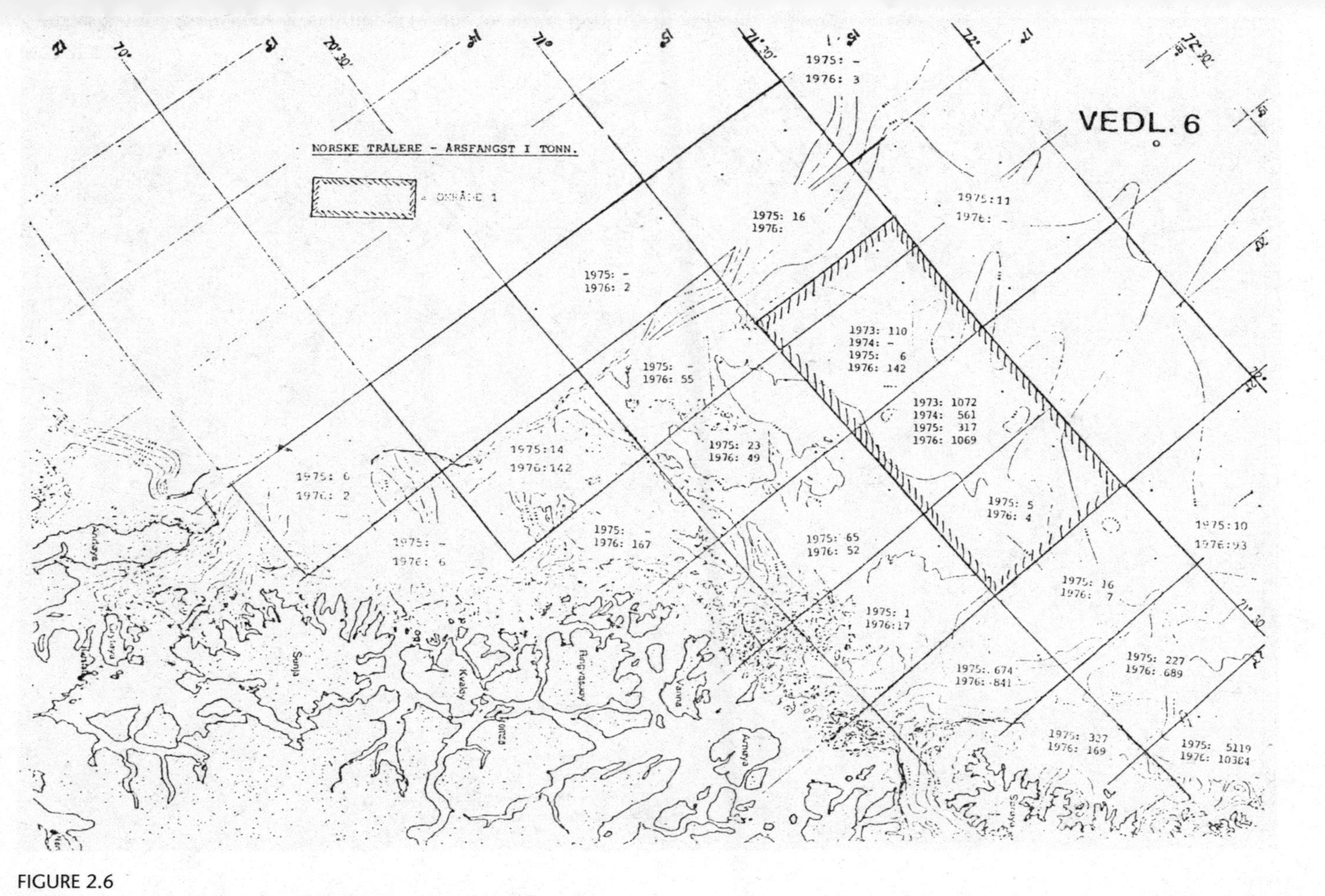

FIGURE 2.6
Catches in tons, from Norwegian trawlers (NOU 1978, 65).

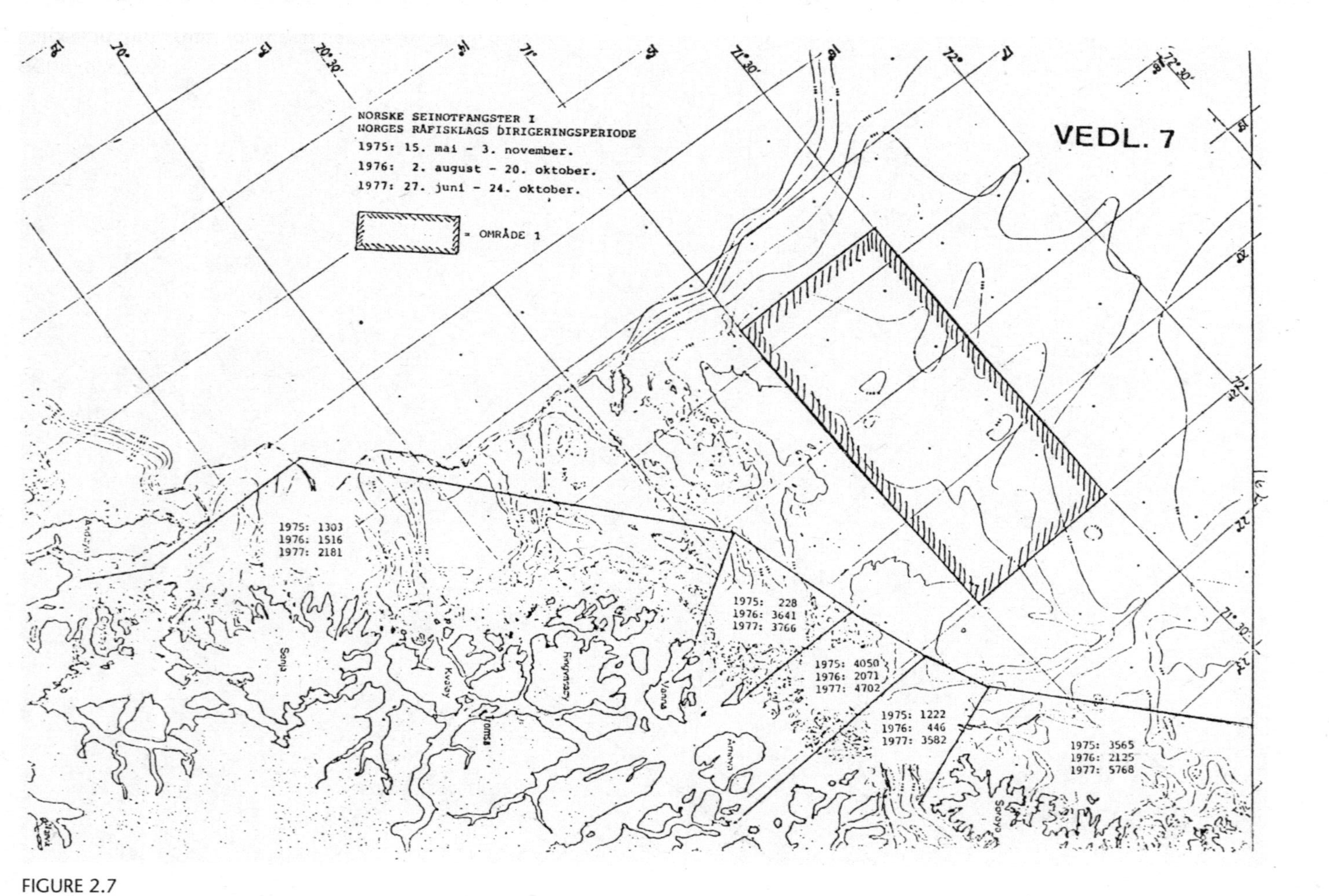

FIGURE 2.7
Catches in tons, from Norwegian drift net fishing for saithe (NOU 1978, 66).

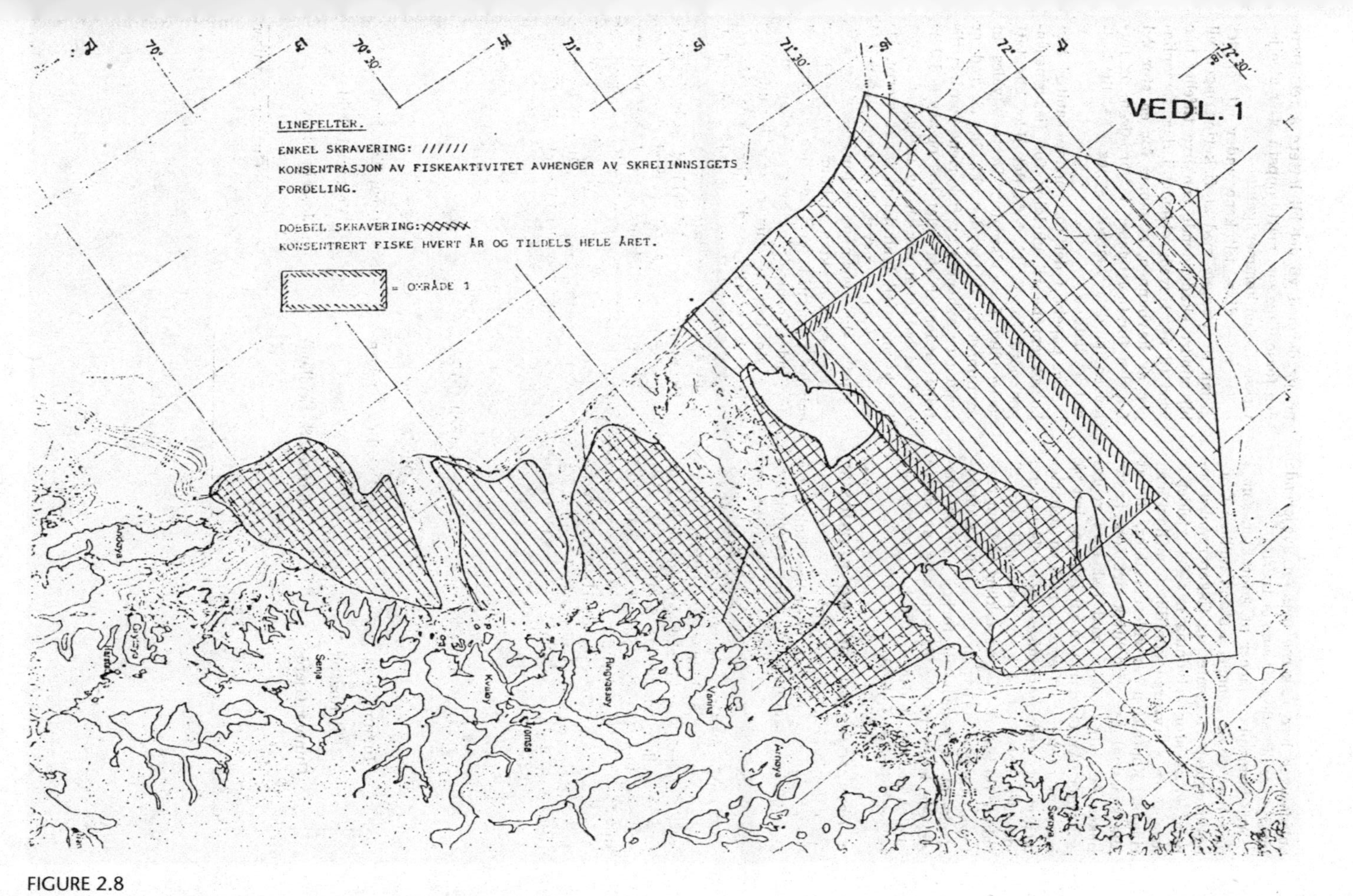

FIGURE 2.8
Fishing fields currently in use by line fishing for cod (NOU 1978, 60).

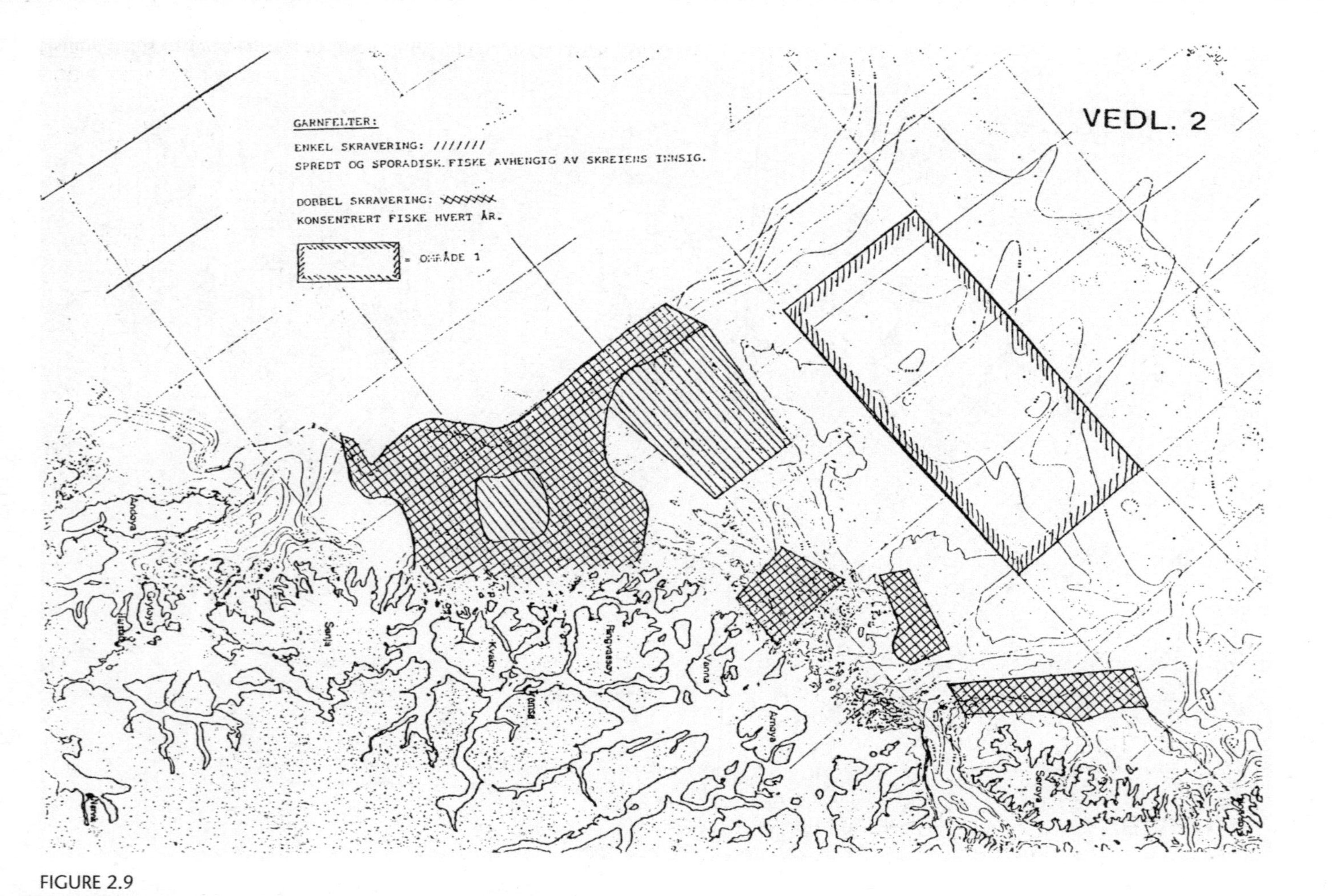

FIGURE 2.9

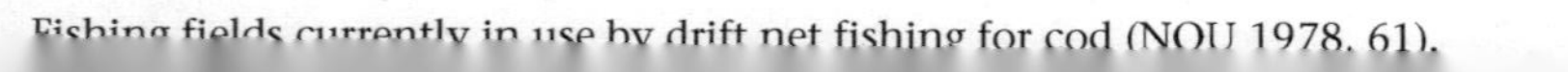
Fishing fields currently in use by drift net fishing for cod (NOU 1978, 61).

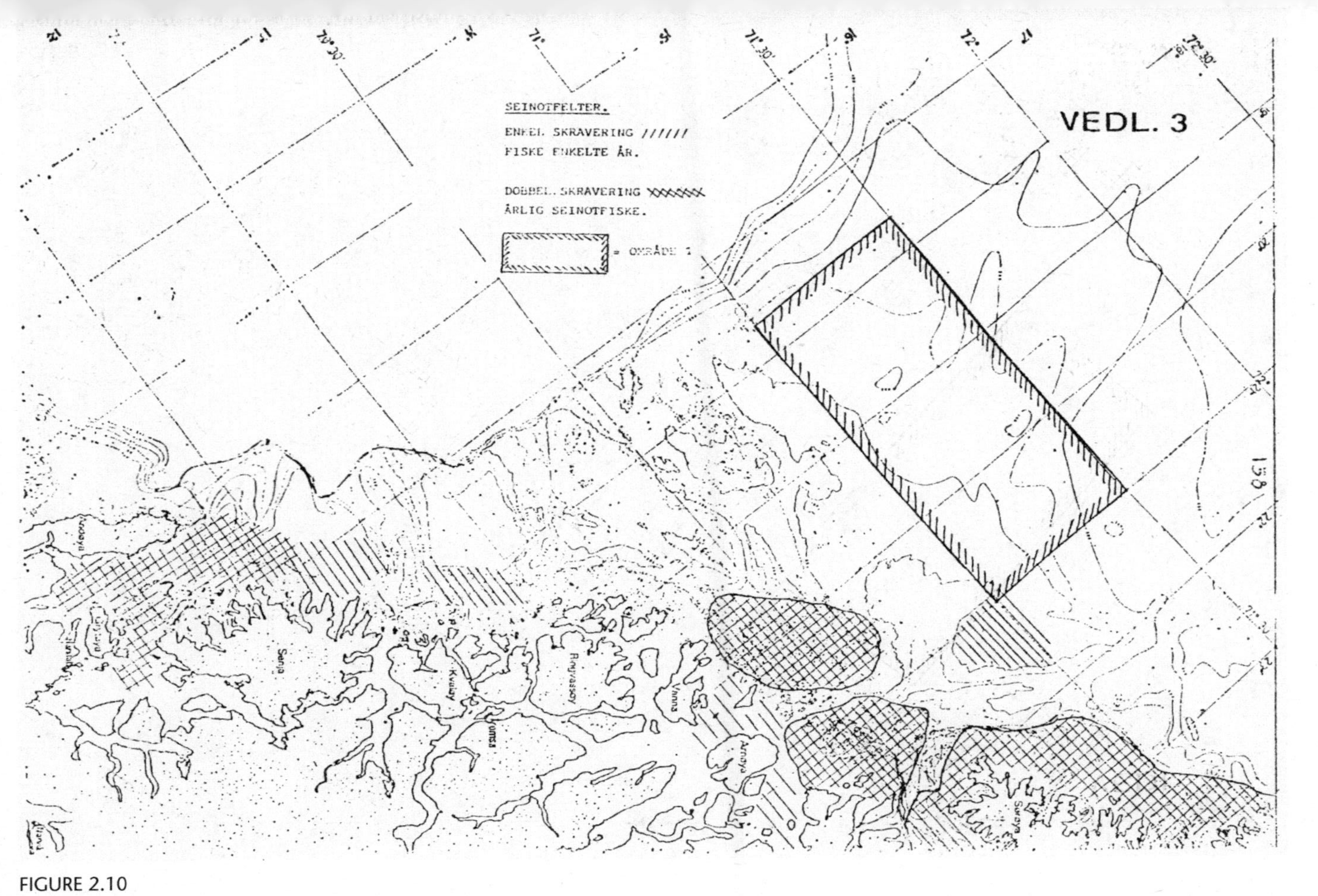

FIGURE 2.10
Fishing fields currently in use by drift net fishing for saithe (NOU 1978, 62).

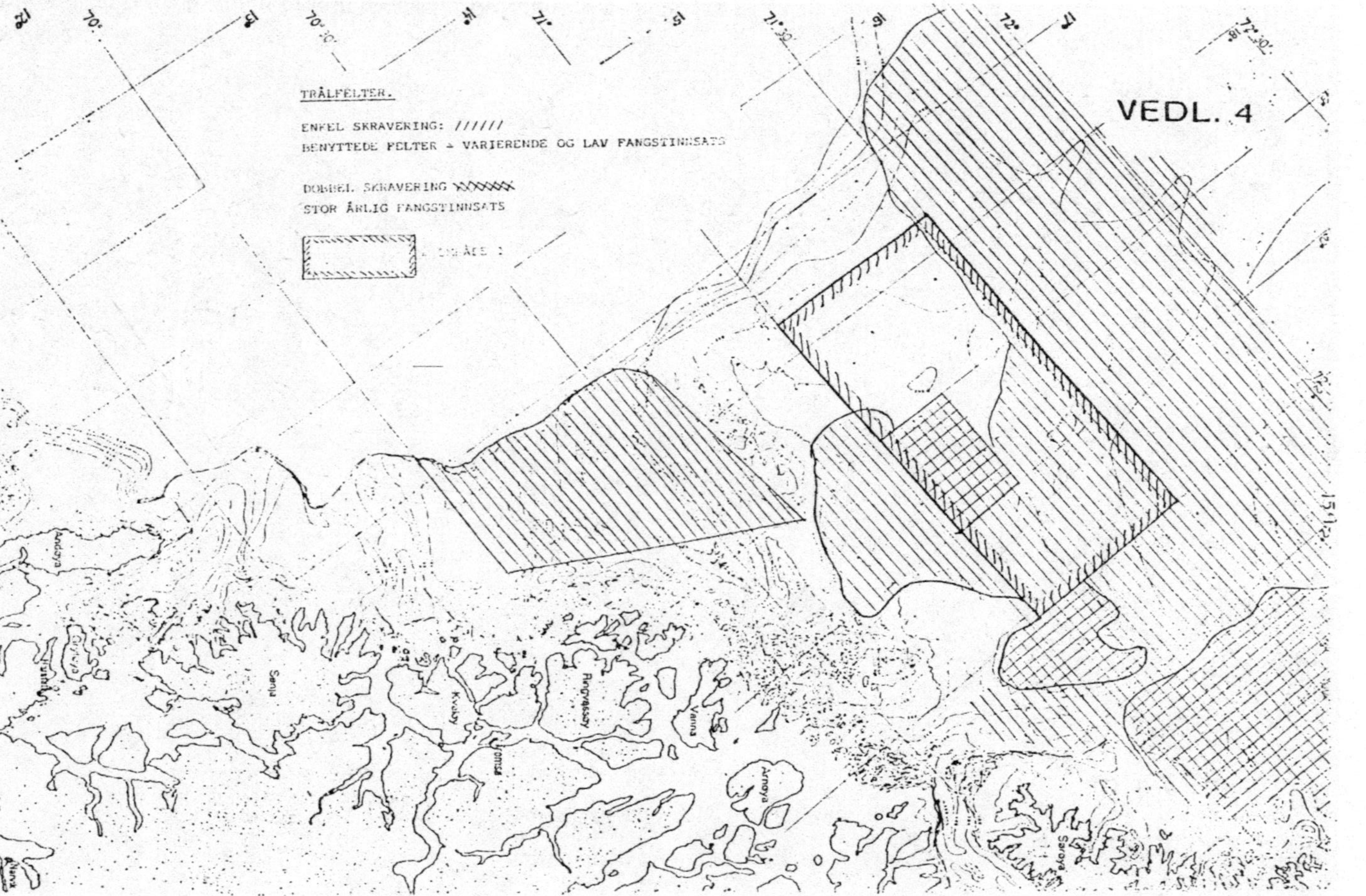

FIGURE 2.11
Fishing fields currently in use by trawlers (NOU 1978, 63).

The fish, however, are not always in the same place. As commented on in the caption of the map depicting line fishing for cod (figure 2.8), the Tromsøflaket fisheries can change from year to year, largely due to how fishers adapt to variations in the timing and patterns of the cod's seasonal influx toward shore. Contrary to the grid structure of the shelf map, which is determined politically and whose lines are set, the fishery maps could only give an indication of where fishing would take place. Inextricably bound to the fish and its whereabouts, the spatiality of the fisheries was not fixed but could be expected to be on the move, with the fish. In the fisheries' version of economizing the ocean, it is not only nature that is being modified; economic practice, too, must adjust. Quite significant to the oil–fish conflict, the maps thereby also stage the rather different spatiality of the capital investments in the two conflicting versions of ocean economization: The petroleum industry's means of production, which are largely fixed on the seabed in the form of pipelines, platforms, and subsurface wells *versus* the fishers' means of production—their vessels and equipment. The latter is a different type of capital. Footloose, or mobile, it is fixed not to place but to the fish.

COUNTER-VALUING BY MAPPING

When compared to the shelf map in figure 2.4, the other maps produced by the Troms fishery manager act as a form of counter-valuing. Inserted with license blocks, but also catch volumes and winding, hand-drawn lines, they let the two versions of economization play out simultaneously. The maps' valuation of the ocean as a space of fixed and mobile capital investments in tension, is very different from the problem-free, win-win vision currently being presented by the OECD and others. Instead, the NOU maps show how different versions of economization come into conflict and give rise to new questions over who has the authority to determine what spatial entitlements different industries should have. The fishers are present, the NOU maps assert, and if the petroleum industry is to move in, they would need, at least in part, to move out. This comes out even more clearly in the assessment made by the Trøndelag region's fishery manager for the Haltenbanken fish bank. Situated south of Tromsøflaket, the Haltenbanken fish bank was similarly being considered for petroleum exploration. Many of the maps depicting Haltenbanken are quite similar to the ones discussed above, but one map stands out, as the Trøndelag fishery manager has here taken the liberty of proposing a *delimiting* of the petroleum exploration area (figure 2.12). This delimited area, the map's caption states, represents a possible *compromise* between the oil and fishery interests.

As emphasized by the Trøndelag fishery manager's map, the NOU enacts the oil–fish conflict as a problem of spatially overlapping and conflicting versions of economization. In describing the potential problems of a northward expansion, the regional representatives of the Ministry of Fisheries had moreover begun to claim a position on behalf of the fishers. They bring out how being on the map, to be known, can be crucial to winning recognition. The ocean, these maps further assert, is not an empty space of "blue" opportunity, neatly parceled and organized in the blocks of gridded lines. It is a seascape of already existing economies—economies that demand to be considered when opening this seascape to new interests and actors. The maps are tools that value by creating a presence—a presence that is quantified by numbers and qualified by the winding lines of fishers' and fish's movements. By situating the fisheries and the values they create, the maps partake in efforts to order the future geographies of the two versions of economization in play. For as the one map produced by the Trøndelag fishery manager suggests quite strongly, the fishers were not simply asking to be compensated for losses and damages already inflicted on them. They were also making claims to the ocean that were of a much more strategic nature, including that of having a seat at the table when the new geographies of the ocean are being drawn. Articulating the fishers' risk of displacement, a claim to staying put was put forward. On the difficult question of compensation, however, both the fishery managers and the NOU are much less clear.

A PROGRAM FOR IMPROVEMENT

Some of the damages suffered by the fishers were easily assessed, as they were covered by tort law. Fishing tools destroyed because of littering, for instance, were already being compensated for. The question of the key value at stake—fish and fishing grounds—was much trickier. It is uncertain, the NOU notes, whether existing law would even apply "when the damages are inflicted upon the fish in the ocean or when it is the actual practice of fishery that is hindered or made difficult" (NOU 1978, 39). For unlike

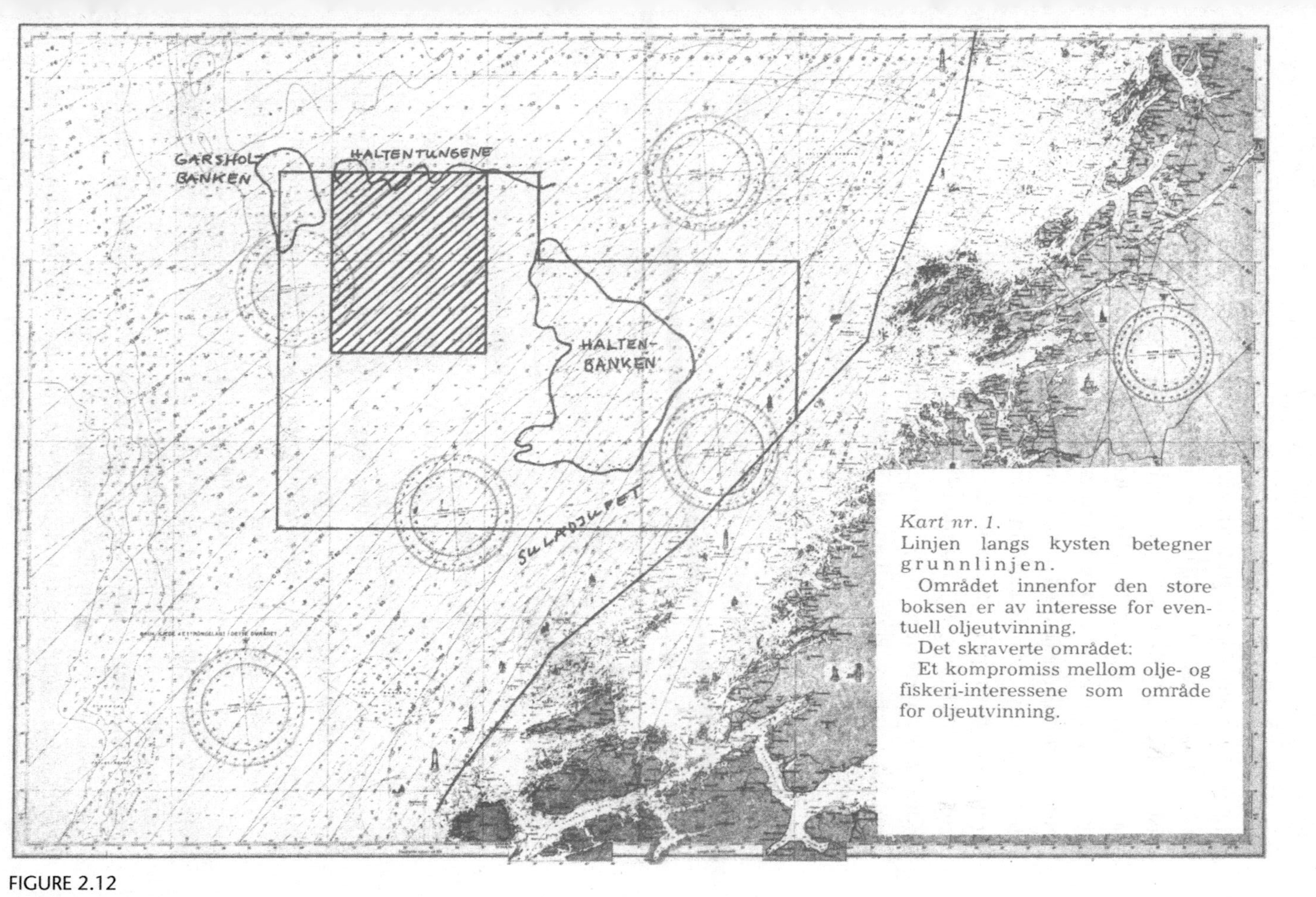

FIGURE 2.12
Trøndelag fishery manager compromise (St. meld. 1978–1979, 117).

the fisher's vessel, gear, or catch, which were protected by private property rights, the loss of fishing grounds was linked to a public right, the so-called *allemannsrett*—"all men's right." "A public right," the NOU (1978, 39) establishes, "is a right that one exercises independent of property rights or any other exclusive rights." In current legal usage there were hardly any cases that could set precedence if a legal party were to seek compensation for its public rights being limited or taken away. Even when it concerns economic losses suffered while engaged in economic activity, the NOU remarks, the courts had been quite restrictive in giving such interests the protection of tort law (NOU 1978, 40). For the fishers to be successful in court, the law would most likely need to be changed, which would be a time-consuming and risky strategy. Rather than risk a prolonged legal battle or the setting of a new precedent, the NOU recommends, it would be desirable to seek a "swift solution." What this solution could be, is not settled before the oil–fish conflict is moved into yet another document—a so-called white paper. Different from the NOU, the white paper is prepared by the government and explicitly for the purpose of presenting matters to Parliament.[15] The white paper, and the subsequent discussion of it in Parliament, often form the basis of a draft resolution or bill to be presented at a later stage. The white paper thereby moved the oil–fish conflict and the issue of compensation into the space of Parliament.

As indicated by the title of the white paper—"Petroleum explorations north of 62°N"—the oil–fish conflict entered Parliament as an issue to be settled in line with the question of the petroleum industry's expansion. Notably, it is predominantly oriented toward explaining why this expansion should now be possible. The white paper notes that the government has put "considerable attention toward improving the relationship between the petroleum and fishery industries" and to "reduc[ing] the conflict of interest between the old and new users of the ocean" (St. meld. 1978–1979, 14). Dialogue with the Fishermen's Association, it is noted, will be a "premise" for any northward expansion (St. meld. 1978–1979, 16). Accordingly, the state had achieved better control, and dialogue with the fisheries had improved. In sum, the white paper brings forth the oil–fish conflict as more or less settled: the state has evolved and is now better equipped to handle the conflicts and tensions of a crowding ocean commons.

Yet, what remained was to settle the complicated issue of compensating for the loss of fishing grounds. Here, the white paper takes its lead from the NOU, conceding to its judgment that settling the issue will raise "very difficult legal questions" (St. meld. 1978–1979, 16). Therefore, it explains: "The Ministry of Oil and Energy has not primarily considered the fishers' demand from a judicial perspective. More important is to find subsidy schemes for the fishery industry that can be carried out in practice" (St. meld. 1978–1979, 16). The white paper then states, without any explanation as to how or why, that rather than compensate the fishers, it will provide them with "support":

> The affected ministries have considered it right to provide support to the fisheries, as the oil activity in certain areas will be a nuisance to the industry, and one has considered which schemes can be relevant. One has decided that a yearly amount in the order of 30 million kroner should be granted to this purpose. How the amount shall be spent will be assessed by the sectoral ministry in collaboration with the Norwegian Fishermen's Association. (St. meld. 1978–1979, 16)

This way of "slipping the law" is the first sign of what soon came to be known as the Oil–Fish Fund. About NOK 30 million was to be spent each year.[16] Fishers could *not* apply for support on the grounds of having experienced loss or damages due to petroleum activities, nor should the fund provide income substitution. Rather, the fund was to support actors with ideas for how to implement "ways of improving the operation conditions" and to develop "alternative operation opportunities" (Innst. S. nr. 327, 1979–1980, 1). Consequently, the issue of compensation was no longer about the difficulties experienced by the fishers, but about improving the ways in which the fisheries were conducted, "converting and adjusting" the fisheries "with the aim of improving profitability" (Innst. S. nr. 327, 1979–1980, 1). Instead of acknowledging the ways in which the fishers operated, the Oil–Fish Fund set the stage for a program of improvement—improving not the conditions of the fisheries, but the fisheries themselves. Nor was the fund set up to work exclusively for the North Sea fishers. It could be allocated to projects both north and south of latitude 62° north, a prerequisite being that improvements achieved through the individual grant projects could, potentially, create new solutions for the entire industry. As it was put in the mandate given to the Oil–Fish Fund by Parliament—it was to "stimulate a best possible exploitation of marine resources" (St. prp. 1979–1980, 1). Curiously, when this mandate was made official in the form of a proposition to Parliament, the amount of NOK 30 million had been increased—again without any justification—to NOK 35 million.

In seeking to settle the oil–fish conflict, the explicit reasoning and justification that Fourcade (2011) points to as one of the key dimensions in legal procedure, when settling compensation claims, is completely left out. In the long run, it was reasoned, a legal settlement in favor of the fishers could also create precedence for suits pertaining to other commons. By keeping the compensation question outside the law and the courts, this could be avoided, at least for the time being. Instead, conflict could be settled in the newly established space of "dialogue" between the state and the fisheries. The making of this space, however, is not specified in the white paper, and so it would be difficult for the fishers and their representatives to hold the government accountable to it. This is also reflected in how the type of compensation given was not actual compensation, but rather a form of generic "support" available to fishers keen on "improvement." The notion of compensation was thereby modified and transformed into something else, and new. Not legal procedure, moreover, but political procedure—most prominently, by way of established methods for producing policy documents and transforming issues—and creating new policies was what the state offered as a means for settling the oil–fish conflict. By the time the issue reaches the white paper, however, it appears that resolving the conflict is no longer the main priority. Rather, the aim is to enable the state's ambitions to move forward and up north with the petroleum industry, the state now being equipped, in part, with a fund whose size appears to be rather arbitrary and whose mandate was to improve the fisheries.

Importantly, this was not a clear-cut top-down or expert-driven procedure. As coproducers of the NOU, the fishers' organizations had been given influence over the very formation of the oil–fish conflict, and they had consented to the nonjudicial, non-compensation solution put forward. The fishers' representatives had also been directly involved in establishing the routines that were now being put in place to improve dialogue about the placement and pace of petroleum activities. The state and its various arrangements—conducting assessments, bringing these into Parliament by way of a white paper, moving the issue to the ministries to be handled—took the users of the commons and their collective concerns into account and was consequently shaped by their concerns. And still, while the government saw it as necessary to *consult* with the fishers, the government did not want them to take control over the decision-making process, drawing maps and articulating compromises for where—and where not—petroleum activities could take place. Direct compensation to the fishers for the loss of fishing grounds would, implicitly at least, grant them some type of ownership or entitlement to specific areas of the ocean, which in turn could impact the state's control over the ocean and its future economization. How the fishers' loss was valued, in the form of the Oil–Fish Fund, was with this closely related to value orderings being put into place by the state to govern the nation's radically revalued ocean.

ORDER AT SEA

Becoming an object of vested national interest, the ocean emerged in the 1960s and 1970s as a space where different versions of economization meet and come into conflict with one another. Indeed, these are tensions that follow Norwegian politics to this day, the waters around Svalbard now emerging as a possible battleground (Steinberg and Kristoffersen 2017), but also the North Sea is being contested anew, as offshore windmills and huge aquaculture installations are set up to replace the platform economy of the petroleum industry. The oil–fish conflict also continued to be an issue handled within the political system, with yet another NOU on the issue being published already in 1986 (NOU, 1986). In what follows, however, we turn from the oil–fish conflict to examine how, for the cod, the ocean is also a habitat. In doing so, we follow the money from the Oil–Fish Fund into activities that were, we imagine, quite different from what the fishers had in mind when making their claims for compensation. For while most of the fund's allocations went to making improvements to fishing vessels, a few interesting exceptions were made for the still quite young aquaculture industry. Heralding the coming of yet another industry to crowd the ocean commons, this brings us to a different version of economization where the economic and nature are put into play: the co-modification of capital and biology toward the constitution of the domesticated cod fish as biocapital.

23
75mm
X

3 BIOCAPITALIZATION: REARING, GROWING, NURTURING

> July 10. After I, yesterday and today, in quiet and clear sunny weather, with the greatest attention, have observed the life and conduct of the young cod, I have arrived at the result, that they attack, kill, and eat one another at quite a substantial scale.
>
> —Gunder Mathiesen Dannevig, quoted in Øiestad 1990, 2

These words were written more than a hundred years ago, in 1886, by the sailor and scientist Gunder Mathiesen Dannevig. The founder of one of the world's first hatcheries for saltwater fish, later to become the Flødevigen Biological Station, Dannevig was a strong believer in the economic potential of putting artificially hatched cod juveniles into the ocean (Store Norske Leksikon, n.d.b). This, he held, could increase cod stocks and thereby improve the fisheries. Still, having peered into the concrete pool where thousands of cod juveniles were swimming about, he found these fish to be both fierce and cannibalistic—an observation that can be read today as a foreboding of cod troubles to come. For despite ardent efforts to domesticate this creature of the ocean, the cod has proved exceedingly difficult to domesticate. Nicknamed the "Houdini of the Sea" for its ability to escape from net pens (Enoksen 2017), the cod has resisted becoming a farm animal in more than one way (Asdal 2015b).

The troubles of cod domestication are also the topic of this chapter, where we trace a series of experiments to domesticate not only this fish, but also its habitat, the ocean. Conducted by marine scientists in the 1970s and 1980s, these cod farming experiments started out at the Flødevigen Biological Station, in the very south of Norway, and were continued on the west coast, at the Austevoll Marine Aquaculture Station. The scientists experimented with different methods for rearing laboratory-hatched cod,[1] as well as with different ways of turning ocean environments into architectures and sites of domestication. If cod juveniles could be produced in large enough amounts and at low enough costs, the hope was, cod farming could be made profitable. Experimenting not for the sake of science only, the scientists attempted to bring the cod into what we can identify as yet another version of ocean economization: aquaculture. Rather than subsea pockets of petroleum or good fishing grounds, this version takes as one of its main resources the reproductive capacities of the living. It reorders and revalues by making the ocean a place not of extraction but of inserting, cultivating, and growing the living, crowding the ocean commons in the ways of domestication. Tracing the cod farming experiments at Flødevigen and Austevoll, this chapter will furthermore show that aquaculture is a version of economization hinged on constituting domesticated species as "biocapital" and on making this form of capital behave in specific ways.

The chapter develops its arguments in close engagement with the cod farming experiments, tracing these from their very beginning in 1975 to when they were stopped in 1986. We apply a combination of methods aimed at recapturing the very work that went into the experiments, including a careful retracing of the cod farming experiments through the scientific papers written on the experiments; archival material from the Oil–Fish Fund (see chapter 2), which was one of the main funders of the cod farming experiments; interviews with several of the researchers involved in the experiments; and visits to the Flødevigen and Austevoll research stations. This enables us to work back in time, while staying close to the everyday practices involved in the experiments. In short, we conduct a form of ethnography of the past, a past that involved trouble and failure, but also what the scientists themselves define as a "breakthrough" in cod farming. By focusing on the co-modifications inherent to *biocapitalization*, the chapter makes an intervention toward critical bioeconomy studies, emphasizing that studies must scrutinize how also "the bio" enters and potentially troubles capitalization efforts. With this, we warn against assuming too quickly and easily that this crucial component to biocapital does behave as intended. This is also an intervention into domestication studies, which we suggest can be fruitfully opened, once again, to studying the economy and capital relations that domestication practices often depend on and are involved in.

BIOCAPITALIZATION

The concept of biocapital stems from bioeconomy studies and the field's analysis of how biological entities—be they live human tissue (Thompson 2005), a cloned pig (Franklin 2007), or genome sequences (Rajan 2006)—enter into and become part of circuits of investment capital. Many of these studies take the 1980s as their starting point, describing how this decade, in the United States, was characterized by intense conceptual, institutional, and technological creativity in the life sciences and its related disciplines (Cooper 2008; Rajan 2006; Rose 2007). Meanwhile, a close alliance was being forged between elite universities, state-funded research, markets for new technologies, and financial capital. As Edward Yoxen (1981) put it in his writings in the early 1980s, life was becoming a productive force, an idea that has later provoked the coinage of terms such as "lively capital," "promissory capital," and, as we explore in this chapter, "biocapital" (Franklin 2003, 2007; Franklin and Lock 2003; Rajan 2006, 2012; Thompson 2005).

Importantly, the notion of biocapital introduces a distinction between capitalizing on the matters of the living and other forms of capitalization, involving other, differently constituted materialities (e.g., the production of cars or smart phone apps). As applied by Kaushik Sunder Rajan (2006, 2012), biocapital is rather eclectically defined, but is generally used to describe how matters of the living— "the bio"—are constituted as biocapital through the co-productions of science, public institutions, policy makers, and investors. Our focus, however, is not so much on how biocapital is made to circulate across such bodies, but rather on how biological entities are worked upon to take on capital qualities in the first place. It is the very act of transforming something into biocapital that we are interested in capturing. Our concern is with this related to what Fabian Muniesa et al. (2017, 14) describe as "capitalization," and which they take to signify a practical achievement and a form of valuation. Capital, they state, "is not a thing in itself—something that one has or has not—but rather a form of action, a method of control, an act of configuration, an *operation*" (Muniesa et al. 2017, 14; emphasis in original). Still, this chapter shows, when linked to "the bio" the transformative act of capitalization becomes more than an action on or configuration of the object of capitalization. It is also shaped by that very object, and by how the living acts on and shapes efforts to constitute biocapital. Domestication processes work upon species to make them behave like a form of capital, but the species that are subjected to such *biocapitalization* efforts can also shape, enhance, resist, or fare badly in such endeavors. Sometimes, they simply escape from them. Our notion of biocapitalization thereby speaks to what we address in this book as the co-modification of nature and economy.

The living—in our case, a cod undergoing experimental efforts to domesticate it—can be troublesome and offer resistance, but it can also act in ways that work with and contribute to its constitution as biocapital. In analyzing biocapital, we must therefore address the twofold process where that which is *constituted as* biocapital is also *constitutive of* biocapital. A species can be made to perform as biocapital, but on the other end, the operations of biocapitalization must often respond to and are thereby shaped by a species' affordances and behaviors. This, then, is what we suggest thinking of as "the behavioral aspects of biocapital." Moreover, and which is made quite clear in our examination, the agency of the cod is not isolated to the affordances of this one species but is achieved in relation to the ocean environments it depends on. This means that in studying biocapitalization we must be attentive to relationality and co-modification in a broader sense than that of the biological entity being subject to capitalization and include, in our case, its habitat and co-dwellers. How are also these worked on by the actions taken and arrangements put in place to make the living behave in ways that conform to its constitution as biocapital?

DOMESTICATION AND ECONOMY

In the 1980s, the cod was seen as a potential candidate for two types of aquaculture (Larsen 1985; Midling 1990): fish farming or so-called ocean ranching. In fish farming the cod was kept in captivity from the incubation of fertilized eggs through to its slaughter, or small ocean-captured cod were kept and fed in a net pen until they grow large enough to achieve an acceptable market price. In ocean ranching, methods not unlike those advanced by Halvor Dannevig were used. Here, cod juveniles were hatched and reared in captivity before being put into the ocean, with the intention to improve cod stocks and, consequently, the cod fisheries. This alternative looks to the ocean as the cod's pastureland, leaving the cod to hunt for the food it needs to grow. As we turn to at the end of this chapter, a third option was also explored in the cod farming experiments—cod farming *without* a net pen—which underlines that the domestication of a species can take place in many ways and by very different means.

In academic literature—anthropology and archaeology especially—domestication has traditionally been thought of as a unilateral process of bringing species of "the wild" into the home or household, adapting these to serve the needs of humans (Cassidy 2007). Hence the term "domestication," whose Latin roots point both to the *domus*—the house—and to the process of bringing nonhuman species under control or cultivation (Online Etymology Dictionary, n.d.). This long predominant understanding of domestication has served as a powerful trope in projects of imperialism and colonialism (Clark 2003), but it has also existed alongside and in tension with understandings that have emphasized the mutuality, troubles, and contingencies of domestication (Cassidy 2007). In the last two decades or so, the concerns of these latter understandings have been captured by a renewed interest in domestication (Cassidy and Mullin 2007; Lien, Swanson, and Ween 2018). Studies have retraced its complicity in Eurocentric views of cultivation and civilization, criticized earlier understandings for invoking a binary relationship between the social and the natural, and argued for increased attention toward the intricate human–nonhuman relations involved.

From considering domestication as a one-way process where humans take control over, confine, and exploit another species, scholars have increasingly come to see domestication as an open-ended and continual process. It is closely connected to both science and the political (Asdal and Druglitrø 2016; Bjørkdahl and Druglitrø 2016) and may well take place through other materialities and practices than those of the farm or the household. David Anderson et al. (2017), for instance, emphasize the "silenced" architectures and infrastructures of Arctic forms of domestication—like tethers, traps, and enclosures—arguing that this represents an unbounded form of domestication wherein the landscape, not the house, comes to constitute the *domus*. Following this, we take *domus* to designate a more broadly defined domestication site and architecture than that implied by "household," "farm," or "enclosure," thereby including how, for instance, distinct landscape uses or—as in our case—experimental, scientific modifications of the ocean environment can also become sites of domestication.

Previous critiques of domestication studies have largely rejected narratives of human mastery and exploitation, including prevalent ideas on the relationship between domestication and economy. For instance, in what Rebecca Cassidy (2007, 5) points to as one of the major works in the anthropology of domestication, the anthology *The Walking Larder*, Juliet Clutton-Brook (1989, 7) defines domesticated animals as "bred in captivity for purposes of economic profit to a human community that maintains complete mastery over its breeding, organization of territory and food supply." Domestication, Clutton-Brook (1989, 7) further maintains, can happen only when "tamed animals" become "objects of ownership." Contradicting this economic determinism, Cassidy (2007, 6) shows, are studies that contest this view of domestication as unilateral, progressive, and increasingly exploitative. By emphasizing the relationships that arise between humans and other species, scholars have instead argued for the potential of mutuality in domestication. Notions of ownership, property, and control are deemphasized in favor of viewing domestication as a symbiosis between species and as involving cooperation, exchange, and serendipity.

Today, this move toward interspecies relationality can be recognized in a prolific and highly influential literature that spans Donna Haraway's (2003, 2008) close engagement with "companion animals" and Anna Lowenhaupt Tsing's (2015) tracing of the matsutake mushroom "in capitalist ruins," to even including the "becomings" of Atlantic salmon in Norwegian waters (Law and Lien 2013; Lien 2015). In moving away from economic determinism, however, studies have yet to find equally rich ways of addressing how domestication is also, and crucially, about bringing species into economy. For while it is recognized that domestication is often motivated by commercial interests or "capitalism," there is still a lack of work that examines empirically and problematizes analytically how domestication most often also entails specific versions of economization. Notably, this relationship between domestication and economization, comes out most strongly in studies that emphasize commodity chains, such as Tsing's (2015) abovementioned work, Koray Çalışkan's (2010) exploration of cotton markets, John Soluri's (2002, 2005) environmental history of the banana, or Roger Horowitz's (2004) account of how chicken changed from a poultry to a meat product, thereby becoming a food commodity of the everyday. These works meticulously examine the relationship of domesticated species and economy, yet do not explicitly address how the processes of domestication and economization are connected. An opening for doing so, we suggest, is to examine how the processes of biocapitalization and domestication intersect: The cod farming experiments we trace throughout this chapter were strongly motivated by the possibility of creating a cod farming industry, thereby stimulating economic growth. Yet, like how domestication is not unilateral, we do not consider economization as a one-way process, but view it through the lens of co-modification. For as we now move to show, the living, the *domus*, and the economy were in the cod farming experiments worked on simultaneously to constitute, adapt to, and direct the behavior of biocapital.

MAKING BIOCAPITAL BEHAVE

Taking the cod farming experiments and, more broadly, aquaculture as a version of economization as our vantage point implies shifting the empirical focus of biocapital studies, as these have largely focused on the commercialization of frontier life science research, and on the novel objects and arrangements that emerge from this. These are often studied by looking at how their "promissory" properties are turned into financial assets and funneled into investment regimes (Birch 2017; Rajan 2006), as well as by interrogating the oft-troublesome ethics of capitalizing on the living (Cooper and Waldby 2014; Franklin 2003; Franklin and Lock 2003; Thompson 2005; Vora 2015). In the highly influential book *Biocapital: The Constitution of Postgenomic Life*, Rajan (2003, 3) emphasizes the increasingly close connections between universities and corporations, and how new life science commodities signal "a new face, and a new phase, of capitalism." However, one has done more to excavate the new economies emerging alongside life science developments than to describe what Stefan Helmreich calls "the *bio* side of things" (Helmreich 2008, 465; emphasis in original). A notable exception is Sarah Franklin's (2007, 57) account of Dolly, "the first cloned mammal to ever be created from an adult cell" (National Museums Scotland, n.d.). Sheep, Franklin here shows, have been integral to changing the shape of capital, Dolly representing a novel form of "protocapital." Neither fixed, circulating, nor floating capital, Franklin (2007, 47) writes, "she is a kind of capital primordium, or source." In this chapter, we extend such empirical attention toward "the bio" by attending to the behavioral aspects of another, though not as radically prototypical form of biocapital, asking how entities constituted as biocapital can act on and *co-modify* the economies they are enrolled in. As we show in our examinations of the cod farming experiments, the behaviors and affordances of such entities can take many forms. They can be material, biological, environmental, or relational, but most often come to view in situations where species such as the cod make some actions possible, others not.

That biocapital studies have largely been oriented toward novel objects and arrangements means that the more mundane, low-tech, and perhaps not so charismatic entities that are also drawn into new biocapital and bioeconomy relations, such as the cod, have received less empirical attention. Cod farming, we however find, is no less interesting to study in relation to biocapital. Perhaps it is even more so, as the constitution of biocapital here unfolds not only within the relatively ordered confines of laboratories and research departments, but in the less-known, much larger, and immensely complex environments of the ocean. The constitution of biocapital is here shot through with trouble and does in fact often fail, which provides us with an interesting case for exploring how, as much as being a productive force, "life" can also turn *un*productive, even *counter*productive. This speaks, we suggest, to what feminist bioeconomy studies have problematized as "biolabor," a notion that stresses that biological entities are not passive artifacts simply worked upon by economic practices.

The notion of biolabor grows out of studies conducted by feminist scholars highlighting forms of work that have traditionally been naturalized or ignored, a key argument being that this work should be recognized as labor. Feminist studies have moreover worked to destabilize what Sarah Besky and Alex Blanchette (2019, 11) describe as the sanctity of labor by "extending what work means, where it takes place, and where it is made," foregrounding instead the shifting and unstable character of labor. These discussions have also been taken up in feminist science and technology studies—for instance, by Rebecca Herzig and Banu Subramaniam (2017, 104), who build on Melinda Cooper and Catherine Waldby's book *Clinical Labor* (2014) to raise the question of "biological labor." Labor studies, Herzig and Subramaniam suggest, may bring further texture and nuance to the understanding of biocapital by conducting robust analyses of the merging of "the substance and promise of biological materials" with "projects of product-making and profit-seeking" (Helmreich 2008, cited in Herzig and Subramaniam 2017, 104). This issue is also raised by Maan Barua (2016, 726). Examining the commodification of animals in ecotourism, Barua argues that far from being passive or inert, animals represent a nonhuman labor, participate in the processes of production and accumulation, and thereby co-constitute political economies from the outset. Still, as the troubles caused by the cod examined in this chapter will underline, the work of nonhumans is not always best captured by studying their participation in work; their refusal or failure to do so can be equally illuminating (Asdal 2015b). As argued by Naisargi N. Dave (2019, 216), "we know that nature works not because it works, or because of how it works, but because it refuses to work."

In this chapter, we take the notion of biolabor as an opening toward examining how, in the cod farming experiments, the cod worked to enable but also to make difficult—and in some cases altogether impossible—its constitution as a form of biocapital. Yet we do not want to consider the cod through the figure of the "worker" or "laborer," as we find this to invest in it the wrong type of agency. The cod is not a wage collector and it is the cod itself that, if its constitution as biocapital succeeds, will become the commodity. Still, there are instances where it can be recognized as performing work and that alert us to how "the bio" should be approached as an active participant in the operations of capitalization. It is, then, such instances that we now turn to examine.

AN ORIGIN STORY: MYSTERY INSPIRED DREAMS OF MASTERY

The cod farming experiments started in 1975 at the Flødevigen Biological Station. Situated at the southern tip of Norway the station faces Skagerrak, the strait that separates Denmark to the south, Sweden to the east, and Norway to the north. Curious about the survival rates of cod juveniles, the scientists at Flødevigen took much of their inspiration from the marine zoologist Johan Hjort,[2] who in the early twentieth century was commissioned by Parliament to investigate the poor condition of the fisheries. In 1914, Hjort published the book *Vekslingerne i de store fiskerier* (The fluctuations of the great fisheries), intervening in what at the time was a heated controversy over the "mystery" of fish stock fluctuations. How could it be that some years, the ocean was teeming with big, fat cod, and other years it was "black" in their absence, the few cod caught being small and scrawny looking? Putting to use new ideas from the then young natural sciences and conducting several field trips to northern Norway, Hjort arrived at a conclusion that has guided the fishery sciences to this day. The great fluctuations in fish stocks, he argued, were due to natural variations in the environmental conditions that determine the survival rate of the fish larvae being hatched each spring. The condition of a particular generation—or "year class," as it is referred to today—will thereby affect future fisheries. The reason for failed fisheries is explained not by the conditions of the present but by the past recruitment of new fish to the stocks.

Picking up on Hjort's findings, the scientists at Flødevigen set out to test two interrelated hypotheses (Moksness and Øiestad 1984).[3] First, that in an environment free of predators, the survival rate of cod larvae would be much higher than it is in nature. And second, that a laboratory-bred cod could survive in nature. This second hypothesis was formulated in direct opposition to the zoologist John H. S. Blaxter (1976), who held that juvenile fish reared in a laboratory are too naïve to avoid predators and have not learned to catch moving prey. To disprove this, the Flødevigen scientists wanted to tag and release cod reared in a predator-free environment, the idea being that recapture rates could give an indication of whether these laboratory fish could in fact survive in the ocean—an indication with huge ramifications for the viability of future cod farming.

At the Flødevigen Biological Station they had at their disposal a saltwater pool much like the one peered into by Dannevig almost a century earlier, which they now prepared for the rearing of cod (Moksness and Øiestad 1984).[4] Cod eggs were hatched at the station's laboratory, and after five days, 200,000 cod larvae were released into the pool. These were carefully overseen, but still, only 100 of them survived. In 1976 and 1977, however, the scientists managed to raise 4,000 fish each year. In 1977, the cod were left in the pool for five months, where they fed on naturally occurring zooplankton. After these five months, one so-called control group was kept in the pool, and another group, of 371 cod juveniles, was tagged and subsequently released into the ocean outside Flødevigen. Altogether, 38 of the 371 were recaptured. The size of these tagged cod, as well of the control group cod, was then compared to other cod captured in the area, the comparison showing similar growth rates for all three groups.

Both of their hypotheses, the scientists concluded, had been strengthened. The laboratory-hatched, pool-raised cod were *not* too naïve to survive in the ocean. This was, however, not the most significant outcome of the experiments. For alongside the production of new scientific facts, a vision had emerged among the scientists: moving forward, the aim was to develop the scientific methods used in the experiment toward becoming methods in the industrial production of domesticated cod.[5] What had started out at the Flødevigen Biological Station as a regular research project, on the still so mysterious life and survival of cod larvae, had become a question of taking control over and mastering what the scientists had sought to understand, of doing reproduction by new means. The cod's propensity to reproduce and grow, it was hoped, could be worked upon to make cod farming at an industrial scale possible. That the cod's own ability to grow—in numbers and in mass—was taken to be the source of economic growth is also key to why we consider this as a move toward constituting it as biocapital. Not merely a production factor or a raw material, the cod represented an entity that if worked upon right could take on the dynamics of capital accumulation. This process of biocapitalization, however, would require further research and investment, which was to move the cod from the saltwater pools of Flødevigen to yet another experimental domestication site.

RAISING STOCKS OF BIOCAPITAL

In December 1979, the thoughts spurred in Flødevigen came to fruition at the newly established Marine Aquaculture Station in Austevoll, an island municipality not far from the west coast city of Bergen. The research station is a division of the Norwegian Institute of Marine Research and was established in 1978 with the purpose of advancing the institute's research in aquaculture. One of the scientists from Flødevigen had transferred to Austevoll, and together with a group of students he set out to build a new site for cod farming experiments, the Hyltropollen pond.[6] The pond was constructed by damming up the two narrow ends of an oval-shaped inlet in the ocean. A small contribution from the Ministry of Environment paid for the start-up of the project, and by spring 1980, the pond was ready for the first experiments of raising cod larvae to becoming cod juveniles.

The pool at Flødevigen had been 4,000 cubic meters, and they had been able to raise 4,000 cod to fully developed cod juveniles: one cod per cubic meter of water. The hopes for the 60,000-cubic-meter Hyltropollen pond was that it could meet this rate, thus producing 60,000 cod juveniles by summer. For the future, it was hoped, the method could be scaled up to produce millions of cod juveniles each year. Numbers, in other words, were a key concern for the scientists involved. This is also reflected in the research papers produced on the cod experiments, where detailed graphs and charts describe the development of the cod population throughout the course of each year's trial (see, e.g., Kvenseth and Øiestad 1984; Øiestad 1985). Already in the planning phase the imperatives of scale and of producing large stocks enter the ways in which the cod is constituted as a form of biocapital. What had previously been thought of as stocks of fish roaming the ocean is transformed, as the domesticated cod is made to be *livestock*, an aquatic counterpart to the many and older species of agriculture (see, e.g., Wilkie 2010). Not an animal that exists on its own, only to coincidentally be captured in "the wild," the domesticated cod is asked to labor, to reproduce and grow, and to do so as a larger population. In fact, as argued by Franklin (2007), the notion of stocks and livestock alerts us to the agricultural origins of capitalism, the relations between sheep, wool, thread, and spinneries being an early example of how stocks of biocapital were raised to produce the input of industrial manufacturing. As exemplified by Irving Fisher's 1896 article "What Is Capital?," the idea of capital as being raised and existing in stocks has accompanied economic theory since its very conception. In our case, such aspirations move from the land to the sea, becoming an *aqua*culture, thereby also transforming the ocean into a site and architecture of domestication. For the cod to become biocapital this place of "the wild" had to be constituted as a place of production, which then speaks to one of the key logics of what we designate in this book as the great economization of the ocean.

In March 1980, about two million cod larvae were released into the Hyltropollen pond (Kvenseth and Øiestad 1984; Øiestad et al. 1987). The larvae were the offspring of a brood stock of local Atlantic cod, bred in indoor tanks at the research station. Hatched in the laboratory five days before, the tiny cod larvae would have looked a little like tadpoles, their heads disproportionally large to their rapidly wiggling tails. As soon as they were released, though, most of them disappeared, scurrying from the surface down to deeper waters. The dams of the pond prevented the water from flowing in and the cod larvae from flowing out (Øiestad et al. 1987, 40; Midling 1990, 11), giving the roughly three- to six-millimeter-long larvae (Grabowski and Grabowski 2019, 137) a growth environment that both mimics and modifies "the wild." In the constitution of the domesticated cod as a form of biocapital, not only the fish, in the flesh, but also their habitat was drawn into the efforts of making this work, making the ocean a site and architecture of domestication. The great question now, as they were put into this new and experimental *domus* for rearing cod larvae, was, how many would survive? How many of them would grow to become cod juveniles apt for the purposes of an aquaculture industry?

In 1980, of the roughly two million cod larvae that were put into the Hyltropollen pond, 150 survived (Øiestad et al. 1987, 40). By 1982, the scientists working on the experiment had managed to raise that number to 10,000, but they were still far from satisfied. As described by the project's lead scientist, the result of the first year was "a catastrophe" and "a laughing stock" (Øiestad 1990).[7] "In all fairness," he stated, "the results of the second and third years were too." Or, as he wrote in a research report summarizing the efforts: "We were practically throwing scarce R&D-funds out of the window. The positive new thing in 1982, however, was a grant from the newly established Oil–Fish Fund. This grant saved the project" (Øiestad 1990, 4). In the years to come, the Oil–Fish Fund would continue to fund the work at Hyltropollen, acting, in the words of the lead scientist, as a "visionary investor" in the project's efforts (Øiestad 1990, 5). For the cod to become biocapital, this tells us, it had to attract and be aligned with another form of capital—visionary and risk-willing investment capital. The notion of "vision," moreover, points to what Thompson (2005) has characterized as one of the key traits of biocapital: the quality of being "promissory" and future-facing, rather than something whose value can be immediately realized. To link a troublesome fish undergoing experimental efforts to domesticate it with "visionary" investments of monetary capital was an effort to move it in a direction where its qualities of biocapital could be manifested and stabilized, so that, eventually, it could become a source of surplus value.

The idea of research funding as taking on the role of a "visionary investor" is striking, but also descriptive of the strong link that exists between parts of the Norwegian marine sciences and what is today spoken about as innovation economy (see chapter 5). In the 1980s, for the aquaculture sector, this link was only just being forged. That the emerging aquaculture industry was seen as holding promise, however, was underscored by a visit to the Marine Aquaculture Station by the then Crown Prince of Norway in 1985 (Fiskets Gang 1985), the future king's presence signaling that this enterprise had the support of the nation. Encouraged by the growth of the salmon farming industry—which in the 1980s would grow from producing just over 4,000 tons yearly to a production of almost 150,000 tons in 1990—aspirations for the cod were also emerging. Could this fish, the most valuable, but also highly variable resource of the Norwegian fisheries be made to grow by the hand of humans, as a cultured species in environments built and controlled by humans?

For the cod to be a profitable aquaculture fish, the Austevoll scientists emphasized, production costs would need to answer to market prices and, therefore, be quite low. Creating methods to produce juvenile cod was not only about growing cod by the numbers, thereby raising stocks of biocapital. To be a productive force, "life"—in this case, a growing cod juvenile—would also have to be aligned with the monetary costs of its survival and biomass increase. This involves efforts of working with and making the living behave in ways that conform to its constitution as biocapital, a form of co-modification that at Austevoll became a question of the cod's very particular food requirements. The concerns of the scientists thereby shifted, from taking command over the cod's reproduction to its *rearing*—from multiplying cod by the laboratory methods of fertilization and cod egg hatching to finding methods to bring the cod up to become a healthy, preferably fast-growing, farm animal.

To see rearing as an integral part of economies relying on the reproductive capacities of the living is to acknowledge that biocapital is not constituted by the "productive force" of reproduction alone. Also, the constitution of biocapital can be about enabling its accumulation through various practices aimed at bringing up, nursing, and nurturing the biological entities and, further, about making these practices of rearing, of ensuring the cod survival and growth, cost-efficient. This was, in other words, about co-modifying the practices of domestication with the imperative of generating surplus value. In the Austevoll cod experiments, this was to involve not only a domesticated cod juvenile, but also efforts to time the growing of the cod with that of other species residing its new *domus*. This points to yet another layer in this aquaculture version of ocean economization: by transforming the ocean into a *domus*, it not only inserts in it artificially bred cod, but also seeks to instill in the ocean a form of productivity, a capacity to make this new form of biocapital grow.

NOT TAMING, BUT TIMING

Surrounded by low forested hills and protected from the open ocean by rocks and reefs, the Hyltropollen ocean pond looks to be a calm, even serene, place. The day that the laboratory-hatched larvae are released into the ocean pond is, however, also the day that most of them begin to empty their yolk sac, the "on-board food supply" that cod are hatched with (Rose 2019, 2; FishBase, n.d.). Their metamorphosis is still far from complete, but the five-day-old larvae have developed the organs and sensory systems required to capture their own food (Grabowski and Grabowski 2019, 137). So, while they continue to draw on the yolk sac for another nine to eighteen days, using it as a form of backup feed, they now begin to develop their ability to actively prey upon zooplankton. On the day of their release, the Hyltropollen ocean pond is, in other words, about to become the site of a feeding frenzy. It is, however, a one-way frenzy. The cod larvae are a desired food for many of the ocean's other creatures, but due to the efforts of the scientists accompanying the cod to Hyltropollen, none of these is to be found in the pond's water: before releasing the cod larvae, the scientists have added rotenone to the pond, a poison widely used for killing fish. Rotenone evaporates quite fast after being released into water, which means that by the time the cod are released into the Hyltropollen pond, its predators have effectively been killed, but the poison has evaporated. The zooplankton it depends on for survival, however, remain.

The feeding frenzy is hence all about the millions of cod larvae that are now set to have their first meal acquired from outside the yolk-sac, fueling the rapid and dramatic changes of their metamorphosis: as the larvae reach a total length of 5–6 millimeters, the characteristic pigmentation of the cod will begin to show, and as it continues to grow, its fins will develop (Grabowski and Grabowski 2019, 137). The caudal fin is first, then the dorsal and anal fins begin to appear as the larvae reach a total length of 9–10 millimeters. Not long after, the pectoral fins follow, and last, as the larvae reach 10–12 millimeters, the pelvic fins develop. During this phase, the larvae will remain adrift in the water, feeding on plankton, but will, at a length of about 17 millimeters, begin their transformation into so-called juvenile cod (Grabowski and Grabowski 2019, 149). At this stage, the fins are fully developed, and the cod develops teeth and scales. The distinctive single barbel—often referred to as the cod's beard—grows out of its lower jaw. The size of these juvenile cod varies, however, with some "settling" at a total length of 30 millimeters, others at a length of 60 millimeters.

No longer a drifting eater of plankton, the cod larvae have now metamorphosed into a fish that is also a hunter. Unlike their counterparts outside the pond, however, these domesticated fish are provided with food: outside the wooden panel of the largest of the Hyltropollen dams, there is a fine-meshed net, and once the cod are large enough not to slip through it, the wood panels can be removed. The net keeps predators out and the cod fish in, but allows the zooplankton it feeds on to enter with the tidal water (Øiestad et al. 1987, 40).[8] As the cod grow larger, they are also fed with dry pellets consisting of fish meal, shrimp meal, wheat, fat, binder, and vitamins (Øiestad et al. 1987, 40). From the beginning of May—less than two months after their release into the pond—they are to be fed every six minutes, and from the end of June, every ten minutes. If Hyltropollen is a site of a feeding frenzy, it is, apparently, one abundant with "prey." How, then, can it be that only a small fraction of the larvae survives?

Studying the pond's ecosystem and the condition of zooplankton throughout the year, the scientists began to compare this to the survival of the cod larvae. Samples were taken daily and, once a week, during the night. Collecting and studying these data showed them that the cod had been put out before the flowering of the zooplankton. This, as it turns out, is important not only for there being enough food, but also that the food is of the correct size.[9] Contrary to the salmon, which is large enough to eat dry feed pellets once its yolk sac is consumed, the cod depends on nutrition from outside its yolk sac when it is still so small that pellets cannot be used; a pellet sized down to fit the jaws and intestines of a cod larvae is simply too small to contain nutrients. Its so-called start-feed must therefore be live feed, like the type of food it would find in the ocean: zooplankton that largely are planktonic crustaceans, and that eat algae.

To the marine fishes of Norway, the most important crustaceans are the copepods. The largest of the copepods belong to a genus called *Calanus*, of which the most important to the cod is the *Calanus finmarchicus* (also called just copepods, or in Norwegian, *raudåte*). *Calanus finmarchicus* spends the winter in deep waters; in the spring it comes up and spawns. The spawning of the copepods and of the cod is, moreover, timed. It coincides with the spring flowering of the ocean and follows from a flow of events triggered by the change of seasons: after a winter with a lot of movement in the water, the nutrients have made their way up to the surface. When the days get longer, and the light is sufficient, the algae flower. And then along comes *Calanus finmarchicus* to graze. The females spawn, and there is a huge production of copepod larvae—the nauplius. These larvae are the main food for almost all the marine fishes in Norway. Also, and importantly, the growth rate of *Calanus finmarchicus* is just about the same as that of the cod, so as the cod grows, so too does its food.

If the laboratory-hatched cod are released into the pond before the hatching of the nauplii, there is not enough food. Or, conversely, if they were released too late, the larvae would have grown too large. What happened in 1980, 1981, and 1982, the scientists concluded from the samples taken in the pond, was that the cod had been released too early. The water had been emptied of nutrients before the cod larvae had grown large enough for the wood panels to be removed, which left the young cod with only one choice: to eat each other. As observed by Halvor Dannevig in the late nineteenth century, cod are cannibals: "they attack, kill, and eat one another at quite a substantial scale" (Dannevig quoted in Øiestad 1990, 2). Understanding the connection between the nauplii and the cod enabled the group to change its strategies, and in 1983 what the involved scientists today refer to as "the great breakthrough in cod farming" happened. They managed to match the timing for putting out the cod larvae with the hatching of the nauplii. Finally, there was enough food for the cod.

To learn from the ocean and mimic the alignments of nature's own time, it turned out, would be essential to the survival rate of the cod juveniles, and hence to the economy of producing cod for fish farming purposes. The times of production, and therefore also the times at which the farmed cod could potentially reach the market, had to be aligned with the times of the cod, the nauplii, and the ocean. Or put differently, the economic concern with productivity had to be co-modified with the temporality of the wider reproductive capacities of nature. This speaks to the behavioral aspects to biocapital, in that it brings out how the propensities of the cod—its rhythms of growth, its requirements toward food, and its cannibalistic appetites—were affecting biocapitalization efforts, simultaneously being worked upon and modified by the scientists' experimental practices of producing cod juveniles.

CONDITIONING COD, LOWERING COSTS

Quite fitting for the experiments that were to follow the Austevoll "breakthrough," rearing entails not only nurturing and caring for a being. To rear can further mean to educate, cultivate, and train. These are actions that are rarely associated with fish, but was exactly what the Austevoll scientists proceeded to do after their 1983 "breakthrough." Having identified two main obstacles to moving forward, they put forward one answer as the solution to both of their problems: to train the cod to change its behavior. The first of these two obstacles concerned the capturing of the fish. To collect by hand 60,000 still quite small, fervently flapping, shy, and now also quite quick swimmers was both time-consuming and labor-intensive, and therefore also too expensive (Øiestad et al. 1987). The second obstacle was related to food and feeding. For although survival beyond metamorphosis had been solved by timing the release of the cod larvae with the hatching of the nauplii, cannibalism continued to be a problem (Øiestad et al. 1987, 39). The cod simply preferred to eat each other, not the dry feed pellets being offered to them. Again, then, survival in the pond became a matter of size, the larger cod eating the smaller, to the point where about 85 percent of the fish were consumed by the surviving 15 percent. Some 500,000 cod juveniles in April had become a mere 75,000 by July. In response to these two obstacles, the scientists at Austevoll took inspiration from the dog training methods that psychologist Ivan Pavlov developed in the late nineteenth century, as well from methods of fish training advanced by professor in technical cybernetics Jens Glad Balchen (Midling 1990). If they could train the cod to enter the collection tanks themselves and to eat the feed given to them, the idea was, this would solve their problems. This training was further to take the form of "conditioning," a method developed in response to observations of the cod's behavior in the Hyltropollen pond and with inspiration from what the scientists described as "recent progress" in conditioning and controlling fish behavior (Øiestad et al. 1987, 2). This effort to condition the cod speaks directly to how the behavioral aspects of biocapital are worked on in domestication practices, with distinct aspects of the cod's behavior tentatively being changed toward a more cost-efficient production.

The scientists had observed that when the wood panels of the dam were removed, the cod had adapted the behavior of stemming the stream and feeding from the water flowing in (Øiestad et al. 1987, 39). The idea was that if this situation were replicated, one could lure the cod to eat the dry pellets, and not one another, by inserting them into artificial currents made from propellers. Mimicking the tidal flow of zooplankton, the propellers were to disperse the feed in the currents. Then, if one added to these currents of food a sound signal, literally calling upon the cod to eat, the cod could be trained to seek out not only the currents, but also the sound signal to find food. Consequently, five propellers were placed in the pond and connected to an automatic feeding system releasing pellets into the propeller streams at regular intervals (Øiestad et al. 1987, 40–41). Thirty seconds before the feeding commenced and lasting until 60 seconds after the release of the pellets, a sound signal created by an underwater loudspeaker would pulse through the water at 15 hertz. Putting into place an underwater camera that could be shifted between positions and depths of the pond, and several transducers and echo sounders that could identify the vertical and horizontal distribution of the cod in the propeller currents, the scientists proceeded to study the cod's reaction to the sound (Øiestad et al. 1987, 40–42). Did they respond? And to what degree were they disturbed by other sounds, like those made by sea birds or boats?

When the experiment started, in mid-May 1985, many of the cod juveniles were stemming the stream of the incoming tide and grazing on the zooplankton. In June, their numbers began to decrease, and, finally, they almost disappeared from the tidal stream. At the same time, the number of cod observed in the propeller streams was increasing. The cod would stay close to the feeding sites and were observed as responding to the sound feeding signal, their appetite being at its peak at dusk and dawn. The cod, the scientists concluded, had been conditioned to respond to and move toward the sound signal; they consistently showed the same behavioral response every time the signal was given (Midling 1990, 7). The next step, then, was to use this new propensity of the cod to lure them into a trap, thereby avoiding the cost of time and labor previously spent on capturing the pond-reared cod. Not the scientists, but the cod—enabled by their new and conditioned behavior and by the apparatuses put in place to support and direct their conditioning—were to do the work.

Having built a fish trap consisting of a two-cubic-meter chamber and two entrances, both of which could be closed, the scientists lowered the trap into the pond (Øiestad et al. 1987, 42). A sound signal identical to

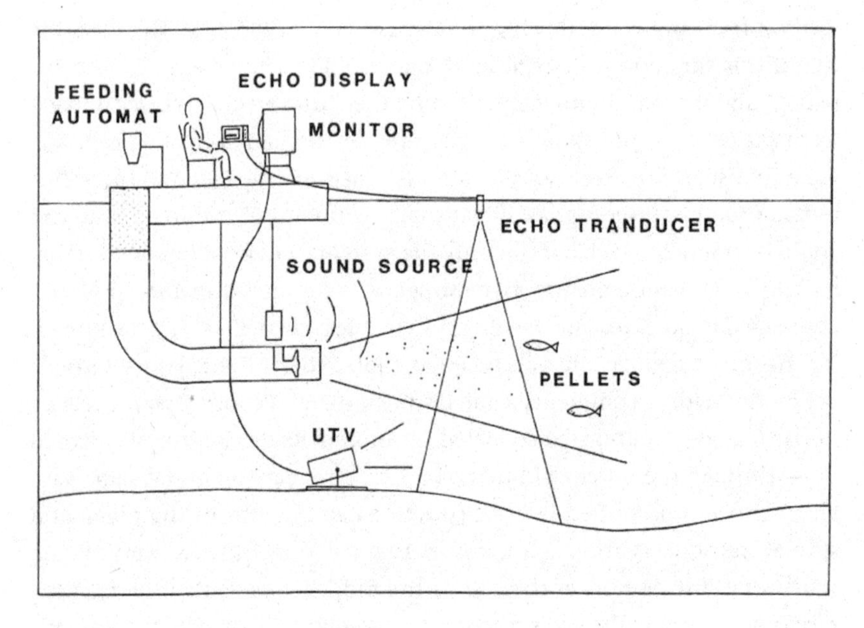

FIGURE 3.1
Illustration published by Øiestad et al. (1987, 40), with the caption "A principal view of the feeding and conditioning system. The pellet was introduced in the stream from a propeller, and in advance of each feeding, a sound signal was given. The behavior of the juveniles was observed by UTV [underwater television] and echo-equipment."

the feeding signal was sent from inside the trap, and while the cod were at first reluctant to enter the trap, they soon caught on, the trap sweeping up as many as 2,000–3,000 cod in one "catch." Once the cod were inside, the entrances were closed, and the cod juveniles were pumped from the trap into a storage tank where they were vaccinated. From here, they were released into a grading tank, where a grid structure was used to sort the juveniles into three different size categories. A "total solution for cod juvenile production" had now been developed, the Austevoll scientists claimed, making it possible to "run this operation on a commercial basis" (Øiestad 1985, 14). Again, we see the intimate connection between, on the one hand, the cod, its propensities, and its behavior (now altered by means of conditioning) and, on the other, its value as a promising species of aquaculture. The scientists' claim that their production methods could now be commercialized further signals how successfully working on and altering the domesticated cod's behavior was seen as constituting a more valuable

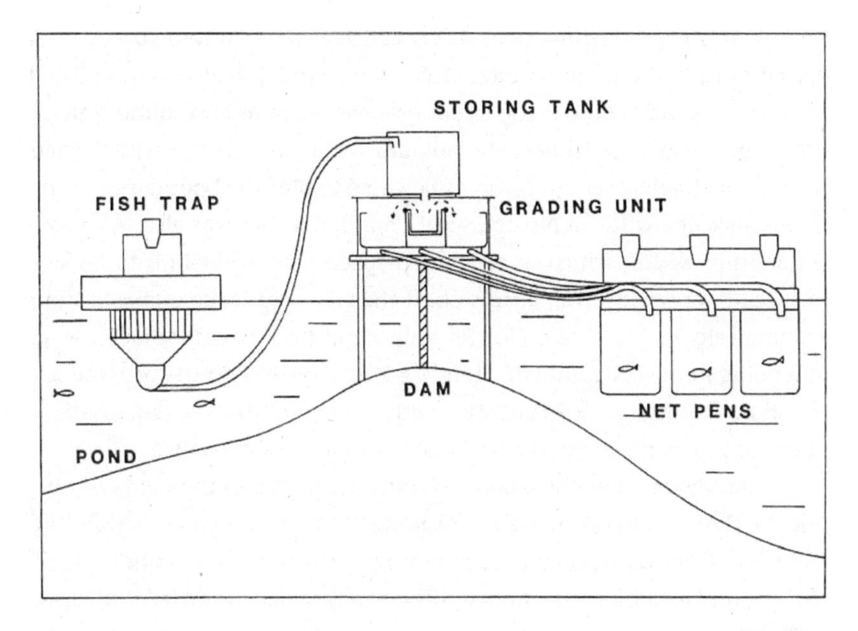

FIGURE 3.2
Illustration published by Øiestad et al. (1987, 42), with the caption "A principal view of the automatic cod harvesting system. The juveniles were attracted to the fish trap by a sound signal in combination with food, the venetian blinds were closed, and the juveniles were pumped to a storing tank followed by grading in three size-groups and then drained to separate sea cages. The process was controlled by an IPLC [Industrial Programmable Logic Control]."

form of biocapital—one whose accumulation was achieved by more efficient means and at lower costs. But what to do with this abundance of trained cod juveniles? Could, perhaps, the scientists were now asking themselves, the productivity of the cod be improved by removing yet another cost—the material structure of the *domus*?

SURPRISES OF A *DOMUS* UNBOUND

As explained above, the cod was in the mid-1980s seen as a candidate for two different types of aquaculture: intensive fish farming or ocean ranching (Midling 1990; Larsen 1985). Combining these two concepts, the scientists at Austevoll moved to experiment with a third way: cod farming without a net pen. Cod juveniles from the pond experiments were to be used in this experiment—about 4,000 cod had been moved to net pens placed outside the pond (Øiestad 1985, 15; Midling 1990). Kept and fed over winter, albeit without the use of propeller streams and sound signals, these latter cod were to become laboratory animals "in the wild." Their rearing was thereby set up in ways that came to blur the boundaries of the *domus* and "the wild," a disruption of boundaries that was also to expose yet another way in which domesticated species are worked on to behave in specific ways. However, at this point the efforts to make domestication economically viable were no longer only about finding cost-efficient ways of tending to the cod's growth. From working with and on the cod, the scientists now shifted to experimenting with ways of putting the environment of the cod to work, an effort that came with surprises of its own.

In mid-March 1986, the 4,000 cod juveniles placed in the net pens outside the Hyltropollen pond were transported to a bay not far away (Midling 1990, 25). Once there, they were put into a net pen, to be "reconditioned" to the sound feeding signal. At first, the cod responded to the feeding signal as they did to other sonic disturbances, such as the sound of a passing boat. With hurried motions they scurried to the bottom of the pen, fleeing the danger that the sound was taken to represent. It did not take long, however, before they accustomed themselves to the once familiar sound and to the feed that accompanied it, resuming their habit of being called on to eat. After a three-month training period the scientists got ready to move to the second stage, only to discover that more than half of their subjects had—as it was put in one of their reports—"escaped from the experiment" (Midling et al. 1987, 10). Of the 4,000 cod initially put in the net pen, only 1,584 cod remained (Midling 1990, 25). The experiment continued, nonetheless, now with the bay having been closed off with a 110-meter-long, small-meshed net (Midling et al. 1987, 2). Lowering three of the walls in the net pen, the scientists allowed the cod to leave the net pen and enter the bay, increasing the volume of water available to them by almost 400 times (Midling 1990, 25). For the first time in almost a year, the cod were again in contact with a natural bottom fauna, and they were to learn how to find their feed in the propeller current produced by the scientists. Underwater cameras, echo sounders, and three divers were placed in the bay to observe their reaction.

Having swum calmly over the lowered edges of the net pen, the cod quickly made their way toward the bottom (Midling 1990, 39–40). Feeding was then conducted in a manner similar to that in the Hyltropollen pond,

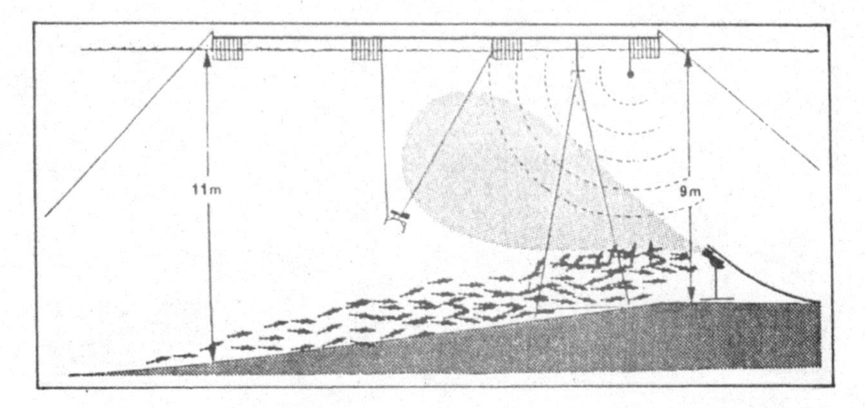

FIGURE 3.3
Illustration of cod behavior (Midling et al. 1987, 6), with the caption "Soon after the sound signal starts, the fish aggregate under the current lobe, waiting for the food."

the feed being released into a current made by a propeller shortly after the sounding of the feeding signal (Midling et al. 1987, 1). The reconditioned cod, however, did not respond. In fact, during the first day, the divers managed to observe only three cod in the entire bay. Two of these were swimming along the net closing off the bay; the third vanished from sight as it managed to wiggle its way under the bottom of the net and escape. After two days, however, small shoals of cod were observed through the underwater camera, and as the feeding regime continued at its regular intervals, more and more cod appeared, swimming closer and closer to the artificial current where the feed was being inserted. Eventually, a pattern settled in the bay: when the feeding signal sounded, the cod began to move about in a large circle. As the shoal reached the point where the feed was to be released, it made a formation at the bottom of the seabed, under the current, where the cod would stay until the feeding commenced. Then, as the pellets were released into the current, the cod closest to the feeding point swam into the current, letting themselves be carried along with it while eating. When reaching the end point of the current, the cod would swim back down, two or three times, and repeat the movement. As the feed would spread through a larger area, the cod formation dissolved, the cod picking pellets out of the water where they could find them.

When the behavior of the cod during and between feeding had been stable for about three weeks, the net closing off the bay was removed (Midling

FIGURE 3.4
Illustration of cod behavior (Midling et al. 1987, 7), with the caption "Immediately after the food arrives, the fish swim into the current lobe and start feeding."

1990, 50). The cod were now free to roam the ocean, and it became quite difficult to establish how many stayed put (Midling et al. 1987, 10). It also became more difficult to determine whether the cod feeding in the propeller stream were part of the experiment or not, as the cod from the pond project appeared to be—as expressed by the scientists—"recruiting" new "clients" (Øiestad 1994, 21); fish who by their own volition had joined the experiment and made use of the facilities installed by the scientists. By the end of the experiment, samples indicated that as many as one in four cod showing up at the feeding point were in fact "wild" cod who had adopted the behavior of the conditioned cod (Midling 1990, 52). Sometimes other species, such as shoals of saithe, appeared at the feeding point, too, and adopted the conditioned cod's behavior toward the sound signal. By introducing a new sound and a new source of food to the bay, the experiment of farming fish without the use of a net pen not only had modified the natural environment of the bay. It was also affecting the behavior of its original inhabitants, who had taken up the behavior of the conditioned cod.

It is not unusual that animals seen as not belonging to the livestock visit the feeding grounds of domesticated animals; some larger fish may even have found the bay an attractive place to eat little cod juveniles. What is quite interesting, however, with these "recruitments"—especially regarding what we describe in this chapter as the co-modification of nature and economy—is that the conditioned cod were now considered by the scientists as performing work—work that, in the case of a future implementation of this method for commercial use, could potentially have an economic up side as well as down side (Midling et al. 1987; Midling 1990; Øiestad 1994). The conditioned cod's ability to recruit "wild" cod was seen as productive, in that it was increasing the stock that could eventually be captured. The conditioned cod, it was remarked, represented "a tidal wave that in the end can lead to *all* cod within an area becoming clients. Then one achieves the favorable situation that the whole cod stock can be harvested in the trap . . . and large, living cod will flow into the market" (Øiestad 1994, 21). Yet, it was noted, if too many cod simply became "clients" reliant on pellets for food, which after all was an expense, rather than utilizing the naturally occurring and, in an economic sense, free resources in the ocean, this could represent an economic challenge. Also, and relating to the other species being attracted, it could become a problem that fish whose market value was far below that of the cod became nonprofitable clients of the feeding grounds. The cod had successfully been put to work—eating and staying put, recruiting and passing on its conditioned behavior to others—but the effects of its labor were somewhat surprising and, perhaps, in the end, not so productive. In these considerations, the ocean begins to resemble a form of arithmetic problem, a place where costs need to be balanced against potential gain.

The absence of a net pen—of a boundary containing the domesticated cod and shutting out its potential recruits—institutes the bay as the *domus*. Like the net pen, it has artificial sounds and feeding machines, but its boundaries are indeterminable. For is it not still, as the appearance of unconditioned, or self-conditioned fish indicates, also a place of "the wild"? Destabilizing the boundaries between the domesticated and "the wild," the experiment of fish farming without net pens brought out not only how the status of the domestic is—as discussed earlier in the chapter—unstable and inherently in a state of becoming. The client cod with its capacities for recruitment also shows that in its boundary-making practices, domestication is dependent on "the wild" being quite stable. "The wild" and the life it contains is not simply something "out there," beyond the boundary of the *domus*, but is instead a *relational* achievement, spun, in our case, by cod fish and other fish; by scientists, a research project, and its experimental practices and apparatuses; and by the ocean, which, with its tides, temperatures, and zooplankton, provided the experiment of "fish farming without a net pen" with a very specific laboratory setting.

DOMESTICATION, BIOCAPITAL, AND CO-MODIFICATION

When constituted through the practices of domestication, biocapital does not act in accordance with some predetermined logic, or simply by the command of humans, but is constituted and unfolds through processes of co-modification. The laboratory-hatched, pond-reared, sound-trained "client" and "recruiter" cod that the cod farming experiments sought to constitute as a form of biocapital came with behaviors and demands of their own. The food had to be the right size at the right time and should preferably be served in moving water. Sounds are scary and make the cod escape, but can also be made familiar and make them come. Working with these behavioral aspects to the constitution of cod as biocapital, the scientists adapted their production methods to the cod, all the time seeking to find ways of making them grow—in numbers and in biomass—while lowering the costs of doing so. This was achieved by understanding the domesticated cod's behavior, but also by finding ways of making the cod work more optimally within the confines of the *domus*. Some did, others escaped, some opted to eat their kin, others to recruit new cod to the experiment, all working in ways as mysterious to us now as the fluctuations of fish stocks were to the scientists of the early twentieth century.

The constitution of biocapital, this tells us, is fragile and unstable. It may well succeed, but it can also fail. Whether or not it does depends on the co-modifications inherent to the act of transforming biological entities into biocapital. There is an interdependence between, on the one hand, the behaviors and actions of such entities and, on the other, the work being done and apparatuses put in place to work on and direct their behavior in specific directions. Also, and as underlined by the hatching of the nauplii as well as of the fish being recruited to the borderless *domus*, to understand the relationship of domestication and biocapital, it is not sufficient to stay with the entities being drawn into new bio-capital relations. As was clearly demonstrated by the experiments of the Hyltropollen pond and the efforts to conduct fish farming without a net pen, the sites and architectures of such activities are crucial to bring into the analysis. If, in other words, we are to understand how nature works, we cannot separate out little bits and parcels of it. We must recognize that by being living beings, the products of an economy hinged on reproduction and biomass growth are also beings with habitats, whether that habitat is a petri dish or, as in our case, the ocean. And the ocean, like the cod, works not *on* command, but by its own command.

Much like how the cod was worked with to constitute its being and behavior as a form of biocapital, the ocean was in the cod experiments configured as a place of production. This is something we also explored in chapter 2, examining how the emergence of the petroleum industry as well as perceived needs for regulation of the fishery commons resulted in a radical reordering of ocean geographies, a reconfiguring that unfolded alongside two distinct versions of ocean economization and using various tools of valuation. In the next chapter, we continue this examination of how the great economization of the ocean has unfolded through the institution of new versions of ocean economization, this time exploring how, with the emergence of aquaculture industries, the coastal waters of the ocean became a space of contested valuations. This time, the conflict was not between two different industries competing over the same areas, but about what we identify as competing "value orderings"—one that argued for modest, small-scale, and locally owned aquaculture, the other for an expansive, capital-intensive growing of the ocean.

21
80 mm
(76)
Skrejunge tagen levende af en Seje Mave.

4 LIMITS TO PUSH, AN OCEAN TO GROW

Entering the speaker's chair of the Norwegian Parliament one November morning in 1985, Labor Party representative Ranveig Frøiland starts out by expressing her warm support of growth in the ocean industries (Parliamentary Proceedings 1985–1986, 626–627). "Industrial growth in coastal Norway," she announces, "is a notion that rings well in the rural areas." She quickly turns, however, to caution against certain types of growth in certain types of industry—aquaculture most prominently. "I want to underline how important it is that this growth happens in controlled forms," the member of Parliament states. The representative wanted not only regulation and policing of aquaculture, but controlled growth also in the meaning of controlling who the new industry was to benefit and who needed to be cared for. The right ties had to be made between the aquaculture industry, local communities, local environments, and the farmed fish. It was the people who lived along the coast who should own, run, and benefit from the industry. And because aquaculture was about dealing with the living, one had to "be aware [. . .]," the member of Parliament said, "These creatures need care and handling in an entirely different way than any other industry product. The larger the facilities get, the larger will also the environmental strains become that lead to disease, stress, and other problems" (Parliamentary Proceedings 1985–1986, 627). Staying small and growing at a modest pace was not only about keeping aquaculture local. It was also about caring for the living. For, as Frøiland underlined, no longer ago than "last weekend," at a fish farm in Austevoll (the very place that we visited in the preceding chapter), between 20,000 and 30,000 farmed salmon had died. Panic broke out among the fish farmers who saw the mass death as coming out of the blue. So, how to care for the living, which was now turning into a form of biocapital in a new and growing version of economization?

Through her address in Parliament, Frøiland raised concern over the pace and place of the aquaculture industry's growth. She did not, however, *introduce* the issue, but rather participated in an already ongoing debate (Hobæk 2023) on how to value this new version of economization. Was modest growth and thinking small the best way forward? Or should one embrace the promises of aquaculture and seek to maximize growth in this emerging industry? At stake was the extent to which one could exploit the ocean as a place for growing both fish and economies. Tensions were emerging with regard to how the living and the economic were to be brought together.

The bringing-together of the economic and the living beings of the ocean now turned into domesticated species, was not only to happen "out there," at experimental domestication sites, such as those described in the previous chapter, or at the new fish farms. As Frøiland's statement alerts us to, these "living creatures" were also to be reared, cared for, and reordered by way of the valuation tools and arrangements of the political.

DOCUMENT SPECIES AND THE REALITIES THEY ENACT

What are the relations between efforts to domesticate a new species, and a state and its machineries? How is it that living species can become part of a state body? What is it that enables a state machinery to act on the cod and other species "out there," in the fjords and in the ocean? And how can we capture this work that adds up to what we have suggested thinking of as the great economization of the ocean?

As noted already in chapter 1, it is tempting to say that large-scale transformations happen due to great forces such as "capitalism," through almost predetermined capitalist "logics," and that transformations must therefore be analyzed by equally large-scale means, such as statistical analysis, large numbers, and vast machineries. And yet state machineries do not work by large measures alone. If we are to understand how great transformations reorder social and natural realities, it does not suffice to follow large measures and large numbers. We also need to follow document work and how documents work as little tools that enact, order, and reorder realities

(Asdal and Reinertsen 2022). Document tools may act locally, but they gain traction and prominence by being simultaneously linked to larger agencies and procedures, budget decisions, manners of deliberation and voting arrangements. Document tools can furthermore work quantitatively as well as qualitatively. They can move issues and concerns, cod lives and visions, into state bodies—as well as out again—in the form of, for instance, budget packages, policy prescriptions, incentives, regulations, and laws. In short, documents can shape and direct not only ideas, but materialities and the living, and when trying to understand them we need to attend to their materiality and their semiotics, their words and their physical properties and affordances, simultaneously.

Documents come in many types, forms, and modes. In other words, there are different "document species." These do not only work as tools that take part in ordering realities. They can also act as "little tools of valuation" (Asdal et. al 2021) involved in efforts to value and ascribe worth, to assess actions worth taking, to praise some things and devalue others, to rank and place things in the preferred order, or what is considered to be the good order—thereby producing patterns and modes of value orderings. This implies that we approach the question of value as *valuations* (Dewey 1939; see also chapter 1), taking this to mean that valuing is something that is actively done and, therefore, that valuations are practical accomplishments (Muniesa 2011). We combine this with an eye for how valuations add up to patterned—that is, *ordered*—realities. In this chapter, we demonstrate the coming into being of two quite different value orderings of the emerging aquaculture enterprise—the first we denote as the "modesty ordering," the second as the value ordering of "ocean growing"—and how these in fact made up not only different, but conflicting versions of economization. Whereas the first was a value ordering oriented toward growing modestly and staying local, the second was oriented toward growing the ocean and growing big. And whereas the first would eventually be turned down, the other was the start of what has later to be turned into what is interchangeably called an "ocean economy," a "blue economy," and sometimes a "bioeconomy." In employing the two key notions *value orderings* and *versions of economization* we emphasize how nature and economies are drawn together and co-modified—yet in quite different ways and with quite different implications for how nature economies are made to transform society.

In pursuing our analysis, the chapter follows a carefully selected set of document species. First, two Official Norwegian Reports or NOUs, a document species already familiar from chapter 2. Second, a publication from an expert group tied to the Research Council for Technology and Science. And third, a white paper presented to Parliament. Together, these documents form a document circuitry where the different documents both speak to and act upon one another. By being, yet in varying degrees and ways, attached to government, they provide access to formal procedures and discourse but also to the issues that they set out to handle, and to the valuation, ordering, and reordering of these issues. Reading them closely allows us to examine the tensions, struggles, and disorder that was also to emerge alongside the new aquaculture industry.

In following the abovementioned documents around, what comes to view is how they are indeed key sites for the economization of the ocean. For if we stay attentive to what these documents do and the realities they enact, rather than being predominantly interested in the positions they represent and what is "behind" them (Hull 2012), it becomes quite clear that the documents are in themselves quite active. They work to draw nature *and* economy together, and to instill a nature economy. This is not to say that this happens smoothly. As already signaled by the worries brought forward in Parliament, conflicts and problems emerge from such encounters. An apt description in this case would be, instead, that there was an issue overflow, a range of different positions and approaches at play, often simultaneously, as well as a seemingly ever-growing series of tensions, disruptions, and, as we will see, disorder.

The document species that we follow around in this chapter are rich entry points for tracing and analyzing *how* it is that nature and economies meet in pursuits to manufacture what we have later been asked to recognize as an ocean economy, a blue economy, and indeed what we in this book suggest thinking of as the great economization of the ocean. Our analysis starts from a situation where these two "things"—nature and the economic—were ordered in the form of two distinct areas of knowledge and practice. And we start out by seeing this ordering of things as an intricate and complex set of issues and problems: On the one hand, the social world of economics and the economy and, on the other, fish biology and the natural world that fish inhabit and swim in—these being, quite concretely, worlds apart. The chapter traces how bodies—government bodies especially—seek to work and act on the economy, nature, and the biological simultaneously, and we take this issue as both our empirical data and analytical problem. We move around with the tensions and conflicts, confusions and troubles that the efforts of bringing these worlds together create. And we take this as an entry point for analyzing the great economization of the ocean; how a new blue bioeconomy version is pushed into place.

While this is an empirical tracing, our procedure is also a scholarly intervention. The last thirty years have seen a series of scholarly interrogations of *the* bioeconomy, and as discussed in chapter 3, a keen interest has been taken in how life is inserted into the economy in unprecedented ways. However, starting from here, there is a risk that we take the bio–economy relation for granted; that we take the "the bio" and "the economy" as already aligned, and as having already joined forces. Instead, this chapter demonstrates the work involved in bringing the bio and the economy together to produce a new version of economization and the troubles that this engenders. We show that the outcome is not only a new version of economization, but crucially also a distinct ocean nature-reality apt to encounter and trouble these large economizations.

AN EMERGENT OCEAN NATURE

The tensions and struggles that was to emerge along the new aquaculture enterprise not only alert us to the scholarly challenge of analyzing how the bio and the economy are brought together, but also to a key problematic in our own vocabulary and theoretical and analytical resources: Across otherwise diverse scholarly fields such as cultural studies, environmental humanities, and science and technology studies, we have become wary of regarding "nature" as a category untouched and uncontaminated by the social, political, and cultural. Concepts and categories signaling interconnectedness and hybridity, such as "nature entanglements," "nature*s*," and "the socionatural," are suggested in its stead and worked with as alternatives and replacements. The very concept of the Anthropocene is effectively a reinforcement of the same turn. Human traces and entanglements in and with nature, this concept suggests, are so intimate and encompassing that singling out nature from such traces and entanglements does not seem to give much meaning.

We share these understandings. It is hard to argue that nature exists in a pure, natural form (Asdal 2003; Haraway 1989). Analytically and empirically, the issue is not, however, as straightforward as one might perhaps believe. Confronted, for instance, with the document material and circuitries that make up the analysis of this chapter, it does not really help much to work from pre-given categories about nature entanglements and the socionatural. First, as we pointed out in the preceding chapter and as will become just as visible in this chapter, the domesticated and "the wild" emerge and play out in tension with one another, creating real problems and challenges in nature as well as in society at large. Moreover, rather than dismiss nature as a category, we need to stay attentive to how distinct versions of nature can be *the outcome* of the practices we study. What this chapter sets out to show is that a distinct nature reality—the nature of the ocean—emerges *as part of* and *in* the very struggles to problematize and counter the pushes and visions for markets and growth. The reality of this ocean nature, we show, is enacted by an ecologically oriented vocabulary. It works through notions such as "ecology," "limits," "thresholds," and "the environment," notions that become inserted into the ongoing economization struggles—not in the form of a towering nature, nature with a capital N, but as entities conjured and mobilized to be part of the issue (Asdal 2008, 2019). Rather than start from entanglements and hybrid categories, we therefore start by attending to how tensions between the living, the natural environment, and the economy play out in the practices that we investigate and we seek to stay agnostic and open regarding which categories to employ.

In making these arguments, we take inspiration from a combined reading of Margaret Schabas and Michel Foucault. In the book *The Natural Origins of Economics* (2005), Schabas traces how economics was "cut loose" from its earlier coexistence with the natural world in a process that unfolded gradually and over the course of the late eighteenth to mid-nineteenth century. Only then, Schabas reasons, "did economic theorists come to see economy as a distinct entity which was not subjected to natural processes, but the human hand and laws" (Schabas 2005, 2). Up until then, economic theorists had regarded the phenomena they studied as part of the same natural world as that studied by natural philosophers. With the changes Schabas describes, the two worlds begin to appear as two distinct fields of knowledge and practice: the natural world, on the one hand, and the economy, on the other. Schabas's work has much in common with Michel Foucault's analysis of how the two disciplines of biology and political economy emerged in the eighteenth century. Organized around the concepts of "life" (for biology) and "production" (for political economy), the two disciplines produced a whole new order of things, Foucault argues in the book whose English title is, exactly, *The Order of Things* ([1970] 1974). Foucault's analysis aims at understanding how these fields of knowledge took part in *disciplining*, forming, the objects of which they speak. Put differently, the words and the objects entered into new relations with one another, which then points to the French title of the same book, *Les mots et les choses* (*Words and Things*). In doing this, the world comes to be arranged and ordered differently, creating new patterns of practice and meaning.

While thoroughly inspired by the abovementioned studies, the ensuing empirical analysis borrows just as much from John Law (1994) and other recent contributions with a more open-ended approach to order—namely, *ordering* as a practice rather than *the* order—and that have an eye for how different versions (Mol 2002) of orderings may overlap and also hang more loosely together (see, e.g., Moser 2008). More than one "grand order," there are orderings. As we will show, such orderings are accompanied by valuations and by value orderings that encounter and come into tension and conflict with one another. The ensuing empirical analysis identifies such *value* orderings in the new aquaculture economy—orderings that, in turn, were decisive to how distinct versions of ocean economization would unfold.

VALUE ORDERING I: JUSTIFYING A NEW ECONOMY BY GROWING SMALL AND STAYING LOCAL

At the time of Frøiland's address in Parliament, a finely woven net of laws and regulations had already been spun around the reproductive capacities of salmon, as well as around other promising species to be modeled on the salmon, like the cod. The new prospective industry was met with keen interest, but also concern: how to spin this net and make sure that the industry, and the species on which it rested, was growing *in a good way*? This emerged as a pressing political issue. Endorsed not primarily for its production of surplus value, aquaculture was instead an economy intended for other types of value creation than those typically associated with business and industry (Hobæk 2023; see also Asdal et al. 2021).

A quite distinct document species had been vital to the speech act situation and the value ordering that Frøiland spoke from in her address to Parliament. This was the genre of documents officially named "Official Norwegian Reports"—or NOUs. This type of document works by excavating what we, with Hannah Knox (2020), can call a "contact zone" between, on the one hand, the state and the relevant ministry and, on the other, an external group of experts, sometimes also stakeholders and interested parties on the relevant issue (see also Krick, Christensen, and Holst 2019). As we have already shown in chapter 2, the custom with this kind of inquiry is to advise the government on how to act on a particular issue and the things at hand. Not only do such documents *underpin* future actions, they also take part in shaping and formatting issues (Asdal 2015) just as much as they ascribe worth, assess actions, consider what is worth doing—and prescribe good manners of proceeding. Consequently, and as shown also by the maps we examine in chapter 2, they both order and value, simultaneously. In short, they are involved in value ordering.

Fish Farming (*Fiskeoppdrett*), as it was simply and soberly titled by the NOU, had been commissioned by the government in 1972 and was returned to government as a finished report all of five years later, in 1977 (NOU 1977). Here, the emerging aquaculture enterprise was ordered according to a form of "modesty ordering" that can be likened to a combination of the famous *Limits to Growth* report, with its call for accepting that there are limits to growth due to limited natural resources (Meadows et al. 1972), and E. F. Schumacher's vision of *Small Is Beautiful* (1973), written in opposition to large-scale production and industry. Concretely, this meant that the aquaculture enterprise was *not* to become a large-scale industry (see also Kolle 2014). Initiated by the NOU, a preliminary law prohibited establishing an aquaculture facility without a concession or a license. This meant that the enterprise was to be ordered by a system of concession rounds—as for the newly established petroleum industry, which we addressed in chapter 2. The objective was for the industry to grow, but the overriding objective was to *not* grow hugely or massively. Small was indeed beautiful, and so was that of remaining local, including being owned locally. In fact, and as demonstrated by Bård Hobæk's (2023) analysis of this early phase of the aquaculture history, the enterprise could only be justified to the extent that it contributed to activity in the local community to which it belonged. The concessions to run aquaculture facilities were to serve these ends: Concessions were to be granted only if new operations were tied to the local community, and areas in particular need of economic activity were to be favored.

This value ordering was quite strictly woven. Not only were the owners of the enterprise required to reside where the fish farming activity was established. The point was also to *not* establish aquaculture as a capital-intensive activity and to *not* have it based on capital inserted into the local community from the outside. Good capital was local and community based, not "big" and external (we pursue this issue of capital qualities further in chapter 5). This was a version of economization coupled with high ambitions of ordering the emerging aquaculture enterprise by use of a fine-meshed net of administrative and political rules and regulations. And the aquaculture enterprise was valued according to its ability to serve specific socioeconomic ends. The relevant question was not so much how to grow, but rather how to make this new economic activity serve specific economic *and* societal ends.

LIMITS TO MARKETS, LIMITS TO BIOCAPITAL

If ever so tightly interwoven with the limits-to-growth approach of the 1970s and a value ordering underpinned by the conviction that the economy was an entity that *could* be ordered—by political means—the modesty ordering of the new economy of fish farming did not end with the 1970s, but continued into the 1980s. In 1985, the circuitry of aquaculture documents was extended, by way of a "sequel" NOU: *Aquaculture in Norway: Status and Prospects for the Future* (*Akvakultur i Norge. Status og framtidsutsikter*) (NOU 1985). And still here, in the mid 1980s, a decade so often characterized by terms such as "deregulation," "neoliberalism," and a "turn to markets," the value ordering was not about setting markets free or growing quickly, but rather about growing only modestly. Yet even more than a concern with limits to growth due to limited *natural* resources, the concern was about limited markets. As stated in the 1985 NOU, "the total aquaculture volume must not increase so fast that the production gets seriously out of sync with the volumes that can possibly be sold" (NOU 1985, 10). Having, as it was put, "unrealistic assumptions" about the speed of the future expansion, one risked "inflicting irreparable damage" on the aquaculture industry (or *næring*, as it is called in Norwegian) (NOU 1985, 10).

The risk and irreparable damage the report was addressing was not so much damage or risk to the "living creatures" or ocean environments in question. The problem was rather that of finding a market in which to place the now domesticated species. As formulated in the NOU, there could be "offset difficulties"—that is, difficulties with selling the cultured species in the market. The expansion had already been positively rapid, sales having grown from 531 tons of salmon and trout in 1971 to about 26,000 tons in 1984—a fifty-fold increase in a bit more than a decade. The expectations for the future surrounding these species, and others to come, were still huge—"so huge," it was noted, that there were "reasons to worry" (NOU 1985, 10). This was not the least due to the fact that more than a thousand applications for a concession to establish a new aquaculture facility had now been submitted. Would it be possible "to cultivate new markets at equal speed to the growth in production" (NOU 1985, 10), it was asked rhetorically. The answer was quite clearly "no": there were limits to markets. The limits-to-growth approach was here tightly linked to an understanding that the size of markets was more or less given. Rather than to grow or modify them, the perceived challenge was to target them with the right form of marketing, in order to have one's share of that market.

Not only were there limits to markets. There were also assumed limits to how well the commodities-to-be, the cultured species, could be expected to perform. How would they respond to becoming *produced* fish, and what scales of production could the ocean take? Such possible limits to ocean nature stood out most remarkably in relation to the species that now, together with the salmon, had the most hopes invested in it: the Atlantic cod. Regarding the latter, the issue was not so much the size of markets "out there," but whether this species could be profitably cultured in the first place. In other words, could the culturing of cod be achieved without involving costs that were too high? As shown in the preceding chapter, the issue turned to that of achieving a well-performing cod biology and to do this at sufficiently low costs. This was about making the cod biology and the economy meet, what we in the preceding chapter investigated through the notions of biocapitalization and co-modification. Here we can see how these concerns were moved, by the document tool of the NOU, to become a concern of the state and public policy.

The question of profitability, however, was a "delimiting factor" (NOU 1985, 11). Not only research had to be invested in the cod, but also patience, as "good profitability" could be expected only "in 5–10 years" of time. Nevertheless, the cod was considered to have great potential—it was a "slumbering giant" (NOU 1985, 12), if one was to believe the most optimistic of researchers, the report noted, yet from a slightly skeptical distance. This possible "giant" resided, however, not in fish farming in the conventional sense. Instead, what the NOU envisioned was that of "releasing fry to 'assist' nature," thereby increasing the cod stocks available to the fisheries and creating what was conceptualized as "culturally conditioned" fishing (NOU 1985, 12). Also here, the question of limits—in nature—came to the fore: one ought not to forget, it is emphasized, that the "primary production" in the ocean is "not without limits" (NOU 1985, 12). The ocean was observed to have a production capacity of its very own. In fact, as we alerted to already above, an ocean nature starts to emerge, assisted by its own specific production vocabulary. The ocean's so-called primary production, the NOU (1985, 12) notes, could be too scarce if cultured cod were released where its wild kin already resided. Most of the ocean's fish populations—no matter whether released or wild, the report stressed—were "competitors for the same food plate" (NOU 1985, 12). The natural environment where the cod was to be reared did not have unlimited conditions for growth. As much as there were limits to markets and so to the profitability a species could yield in production, there were also limits to nature's own production capacity. In this way, two "production systems" were entering into possible competition and conflict with one another. On the one hand, this was nature's own ordering: the system of "primary production" in the ocean and its own so-called "sustenance foundation," both of which were key to the "wild" cod fisheries. On the other hand, there was a societal and *cultured* form of sustenance: the emerging aquaculture enterprise and a cod that was to be cultured and assisted by human and technological means. The latter, the cultured cod, could, if let into nature's own order and production system, end up competing with the wild cod.

As we saw above, the concern was not only about the capacity of the ocean environment and what the ocean could absorb, but just as much about the capacity of the market and what the market could absorb of such cultured commodities. What if this "culture-assisted" cod ended up competing with the cod-in-the-wild, not only at sea, but also in the market? Competition, also price-wise, was envisioned between the two "species" of cultured and wild cod (NOU 1985, 53).

THE PRODUCTION OF NATURE AND ECONOMY

Thinking with Schabas (2005) and being aware of how the discipline of economics has its roots in the natural world, it is instructive to dwell somewhat on the conceptual vocabulary that the above issues were discussed within. For instance, the Norwegian word *næringsgrunnlag*, here translated as "sustenance foundation," can refer to *both* the natural world and the societal phenomena of business and economics. In other words, it is a concept working at the intersection of nature and economy. *Næring*, directly translated, means "feed" or "nourishment," but is also used as a term for sustenance on a larger scale, often taken to refer to economic activity, business, and industry; *grunnlag* means foundation or basis, pointing to what one can build such activity, business, or industry on. The two worlds and phenomena—the natural and the economic—are thereby linked together at a conceptual level, signifying how they are rooted in related problematics, namely, "nourishing" in the meaning of the foundation for making something grow, be it in nature or in the economy. This is not only about how economy is rooted in or exploits the natural world. Rather, this is about how both nature *and* the economy become ordered around the notion of "production." Production, Foucault ([1970] 1974) observed, was the main point of reference for the political economy of Adam Smith ([1776] 2008) and others in the eighteenth century. The economy no longer ordered around exchange or circulation, but instead became ordered around labor and production. Interestingly, what we see here is how nature is ordered around the very same lines, as a site of production in its own right. As the ocean is increasingly becoming a site and object of economization processes, it is also being transformed into a site of production in its own right. As we will see, a production orientation was to become a key point of contestation regarding the growing aquaculture enterprise

Also in other ways, the NOUs on aquaculture in 1977 and 1985, draws nature and economy conceptually together. We can observe this for example when from how economy is being linked to notion of "life," the notion that Foucault observed as the key invention accompanying the emergence of biology. In Foucault's analysis, "life" emerges in the late eighteenth century as an autonomous object of knowledge (Foucault [1970] 1974, 162). In the two NOU's "life" emerges as an object of knowledge *and* site of intervention *across* biology and the economy, and the notion of life informs and shapes what economy is taken to be or become.

The government's mandate for the NOU commissioned in 1972 was for it to inquire into how artificial hatching and culturing of fish could develop into a *levedyktig* form of *næring*—or enterprise. A translation of the word *levedyktig* could well be "viable," but this does not capture the entirety of the word's meaning. More directly, the word translates into "able-to-live," or even "life-skilled." And the mandate with which the commission was equipped was precisely that of inquire into an enterprise that could sustain life, literally a *"life-able"* enterprise (NOU 1977, 7). In this specific context, this signified an activity that was able to sustain an economic life, and so it captures the problem at the heart of the envisioned new economy, namely that of: sustaining natural life and an economic life simultaneously. How could the two be drawn together and made to work together?

In the 1985 NOU, the act of co-modifying nature and economy comes across as a double operation. On the one hand, the operation of drawing biology and economy together so that the propensities and affordances of the species in question are aligned with economic concerns related to costs, ownership, prices, and markets. On the other hand, it is about a mode of ordering nature in opposition to and against demanding and intervening operations of economization. We saw the latter already in the oil–fish conflict that we analyzed in chapter 2, where the assembly of fish and fishers were presented and realized by maps that worked as as a counter-valuation to the emerging petroleum economy. Here, in this circuitry of policy documents on aquaculture, something related happens. The scientific vocabulary of biology and later also ecology, is set to work as a way of realizing and re-presenting an ocean nature, and with this it also acts as a re-valuation of the ocean. The differences between the sites of co-modification, the closely related 1977 NOU and the 1985 NOU were not dramatic. The latter stayed within much of the same value ordering as the former. Yet, their differences alert us to an emerging shift with regard to how a "limits to growth" approach is justified. In the latter, a distinct ocean nature was emerging as a value-order in its own and which, due to its own lack of production capacities, could delimit growth. In this way, an ocean nature was emerging in response to the prospect of a growing aquaculture enterprise. This was a value ordering very different from the version of economization where the aquaculture enterprise was all about societal demands and concerns about staying local, small, and employing resources that belonged locally. Now the delimiting factor to growth was not any longer societal and the demand to stay local, but natural. Moreover, as the 1985 NOU noted, the conflicts one imagined could be expected to multiply with a future crowding of the ocean commons. Conflicts would grow, it was predicted, when the cultivated areas of the ocean increased in volume as well as in value. Like one had earlier planned for conflicting usages of the fish banks far out at sea, one now had to plan for conflicts in the coastal zone (NOU 1985, 83).

The modesty ordering that we have retraced and delineated was however up for contestation. And if we read closely, we can trace how this played out at a conceptual level and how the 1985 NOU through its very wording was involved in defending a modesty ordering. The most prominent example is the contestations over the new concept of *havbruk*, which can be translated as "ocean growing." In Norwegian, the word *bruk* carries a double meaning. It is both to utilize, "use," while also signifying a form of enterprise, most often an enterprise linked to the use of natural resources, for instance, as in *landbruk* (agriculture) and *fiskebruk* (fish processing plant). More than in the Norwegian wording, then, the translation as ocean growing points more directly to how aquaculture was moving from a modesty ordering into a growth and production enterprise. Different from aquaculture, the notion of *havbruk*—ocean growing—opens up the possibility of not only cultivating and farming fish, but of turning the ocean into a farmland and rendering the *ocean* a site for growth and production—a site of expansion and extension. It is precisely this which is being questioned and contested in the 1985 NOU. "The concept of havbruk [ocean growing]," it is noted, "has entered the scene as a buzzword," used "by some" as if pursuing aquaculture was "something radically new" (NOU 1985, 10). Moreover, and due to how this phenomenon was talked about "by some," it was as if there were no limits to the growth of aquaculture, or "as if there are no limits to how extensively it can be developed."

In the NOU outlining of the concept of ocean growing, we are alerted to how the modesty ordering is now being challenged. In fact, the title of the document against which the NOU seems to be commenting and writing was titled, exactly, *Growing the Ocean* (Jensen et al. 1985). Its objective of this document is furthermore to carve out the ocean as a space for blue growth. Considering how its cover is also blue and how its contents are largely oriented toward laying out a blue-print for a new ocean growing economy, we will henceforth refer to this document as "the blue book." Its preface, collectively authored and published in cooperation with the Research Council for Natural Science and Technology, was signed on the very same day that the 1985 NOU was handed over to the Ministry of Fisheries—May 13, 1985. An earlier version of the blue book had already been written and presented to the same research council (Mariussen 1992). When complaining about "Ocean growing" the 1985 NOU's frustration can be traced back to the blue book, its program, and its collective of authors. "Ocean growing" was now being pitted against the more closely related "fish farming" and "aquaculture," the latter two being in the titles of the 1977 and 1985 NOU reports. Two value orderings were at play and in tension with one another, a tension that can be articulated in the following questions: Was that of domesticating fish now to become the basis of a whole new industry—an enterprise where the ocean could be made to grow and become the new farmland and a growth-oriented enterprise? Or was it rather to remain a *culture*, and a local and small-scale one, as signified by the word "aquaculture"?

VALUE ORDERING II: OCEAN GROWING, TENSIONS, AND DISRUPTIONS

How does "growing the ocean" take form as a distinct value ordering and version of economization? A first clue can be found on the very cover of the blue book where—somewhat hidden behind swimming fishes, algae, and mussels—a huge shovel is placed right at the center of the picture, as if digging into the sea bottom (see figure 4.1). A rather blunt parallel is in this way drawn, associating *aqua*culture with *agri*culture. Despite the simple shovel, however, this is not about small-scale agriculture translated into aquaculture, but about aquaculture transformed into an industry, borrowing methods and breeding techniques from agriculture.

The blue book's chapter on breeding is quite telling of how this economization is envisioned. The history of animal husbandry, it begins to highlight, contains a series of convincing examples of active breeding for "an enhancement of the animal material" (Jensen et al. 1985, 56). In an illustration, two different versions of pigs are shown, the first being a boar, whose Norwegian name literally means "wild swine," and the second showing a regular domesticated pig (Jensen et al. 1985, 57). The latter, it is noted, is a

FIGURE 4.1
Illustration from the front cover of *Growing the Ocean: Perspectives on Norwegian Ocean Farming* (Jensen et al. 1985).

descendant of the former. It is a modified version of the boar and represents what is denoted as the newest "model of the year" (Jensen et al. 1985, 57). Here, animals are valued as an input and a "material" with the potential to be modified and given enhanced value. A more *productive* form of life is ostensibly being put forward. Fish is approached in the same way. The fact that the farmed salmon *appeared* very much like the wild version of the same species, the blue book comments, signals "vast possibilities" (Jensen et al. 1985, 57). The salmon, it is even claimed, has a potential for "breeding progress" that greatly outcompetes what is possible in conventional animal husbandry (Jensen et al. 1985, 58). This view of the salmon as great breeding material is furthermore integral to the blue book's overriding value ordering, which, as captured by its title, is "ocean growing."

Ocean growing, the blue book states, implies "a systematic use of the ocean's capacity to produce food and raw materials" (Jensen et al. 1985, 7). The possibilities are taken to be vast. The ocean's prospects are equated with land-based production, the only difference being that the former is not yet taken into productive use. For instance, it is noted that the so-called primary production is about the same size in the ocean as it is on land. And still, on a global basis, humans retrieve only 3 percent of their nutrition from the ocean (Jensen et al. 1985, 11). The major reason for this, the blue book argues, is that while one is systematically *growing* the land, one is only harvesting from the ocean (Jensen et al. 1985, 11).

Norway, the authors of the blue book claim, is particularly well-adapted to and holds the right competence to realize the ambitions of growing the ocean. Moreover, and in contrast to the 1985 NOU, it is envisioned here that this is something the market, internationally, is ready for: "The market potential for ocean grown products must be considered as great," the blue book states. If production could match this potential, it would pave the way for long-term contracts and stable deliveries of high-quality products adapted to the customers' demands. Compared with traditional fish consumption, it is noted, ocean growing would introduce a new dynamic and open up "entirely new market segments" (Jensen et al. 1985, 17). With this, the qualitative aspects to the market were changing too: From being considered in the 1985 NOU as a relatively stable entity, a pre-given site into which products are placed and marketed, the market enacted by the blue book is a much more modifiable entity. New so-called market segments could be carved out, and the products could be modified together with them—co-modified—and aligned with the demands of the consumer. This is a market whose limits to growth were fewer and more amenable. The same applies to the ocean. Rather than limits to ocean growth and to the ocean's own productivity, there was unfulfilled potential. The ocean conjured in the blue book was far from crowded, instead there was abundant space for more—given that the ocean was systematically grown.

Two different and in part opposing versions of economization and value orderings now existed simultaneously: the modesty ordering of aquaculture and the far more expansive value ordering of ocean growing. In line with standard procedure for how NOUs move within and beyond government, the NOU inquiry of 1985 was followed by a round of consultations. Interested and relevant groups of actors were invited to respond, in writing, to the report's judgments and recommendations. Next, and based on this process, and according to standard procedure, a white paper was submitted to Parliament for consideration and deliberation. The white paper then, as also in line with standard procedure, had moved the issue out of the hands of expert groups, and was instead taken further and developed inside government, in this case by the Ministry of Fisheries. With the title "On ocean growing" (St. meld. 1986–1987), the white paper was submitted to Parliament in June 1987.

As we have already stressed, these document species are not simply mirroring events that happen elsewhere, *outside* the documents. Policy documents are in themselves part of actions and sites for ordering action (Asdal 2015a). They are "little tools" for ordering and realizing events and things to become. A white paper presented to Parliament can therefore be considered as a specific form of action and as a site for acting and working on things and issues. In the case of the white paper "On ocean growing," we can sense a change of direction already in the title, as "aquaculture" had been supplanted by, exactly, "ocean growing." Not only was the title advocating a more ambitious and growth-oriented approach; the very situation had changed quite radically, from aquaculture being *promising* to what was now described as "an explosion-like production growth in Norwegian aquaculture" (St. meld. 1986–1987, 126). In 1980, the white paper tells us, the total production of salmon and trout had been about 7,500 tons. In 1985 already, the numbers were significantly higher: 45,000 tons of salmon and 4,200 tons of trout. During 1986 only, the production of salmon had increased by 53 percent. Selling these increased production volumes had not at all been difficult (St. meld. 1986–1987, 126), implying that markets could indeed be grown. However, not only had production and exports been growing, so had the problems, tensions, and challenges.

CARVING OUT AN OCEAN NATURE

Sometimes, laying out a policy document—what it is doing, how it is reasoning, and the ways in which it is moving its objects toward a particular mode of ordering—is particularly challenging. Not the least if we are searching for a form of order, logic, and coherence. The 1987 white paper "On ocean growing" (St. meld. 1986–1987) is a case in point with respect to this problem. In fact, what we find here is not so much an ocean economy logic and order, but *dis*order. When reading the document, following its sections and chapters around, it is as if everything moves in all directions simultaneously: Within the one and the same paper we find efforts to tie the aquaculture industry to a local community; a massive eruption of environmental problems and diseases; a loosening as well as tightening of regulations; an industry outgrowing most of the preceding expectations when it comes to, precisely, growing; pushes for growing more and more rapidly; arguments that there are and must be physical limits; and amid this, a series of troubles and efforts to align the life of fish bodies—in the form of living creatures and specific versions of fish biology—with business, the economy, and markets.

While the white paper works toward optimizing conditions for the continuation of the "explosion like" growth, it strangely also reads as a site for a "limits to growth" approach. The ocean is enacted not simply as a space for growing but also, as we observed already in the 1985 NOU above, as a system in its own right. Here, however, this approach is taken further. More precisely, the "ecosystem" (St. meld. 1986–1987, 17) is brought into the document as a key definition to be considered. With this, the notion of the eco-system works as a tool for establishing the ocean as a particular version of nature. A series of elements intervene toward this: the notion of "carrying capacity" (St. meld. 1986–1987, 16) is introduced and defined as the maximum number of organisms an area can sustain without it becoming permanently deteriorated. An ecosystem, the white paper further explains, is composed of all the nonliving and living factors within a particular area. It can be a small pond, a part of a fjord, or an ocean area, or it can encompass the entire biosphere—that is, the parts of the earth (soil, water, air) where living organisms can exist. The white paper further points out how the different organisms in an ecosystem together add up to a food chain or a food web, and that there will always be the most energy or calories available at the lower levels in the chain (St. meld. 1986–1987, 17). For each level up, there is a huge energy loss. Through definitions such as these, the eco-system vocabulary works very concretely to carve out a nature that demands that one treads carefully and has consideration for these quite specific ecological conditions. Different from the ocean nature to which we were introduced earlier, this is not simply and straightforwardly a site of production and work, but a site with an ecology and, consequently, a system of relationality and interdependencies.

As we saw in chapter 2, the ocean had in the 1960s and 1970s already been reordered and revalued into extensive economic zones. Now, what we can consider as "nature zones," enabled by the concepts and science of ecology, were following suit. Yet ecology, with its attached notions of "carrying capacity," "populations," "biomass," and "genetic variation," did not stand alone. On the contrary, the two radically different value orderings—that of modesty and that of ocean growing—are inserted into the same white paper. Interestingly, this is done without the two value orderings encountering or confronting one another or being explicitly and critically discussed in relation to one another. If both were equally relevant to the enterprise of ocean growing, how were the two to meet and be cared for, *simultaneously*?

LOSS, DEATH, AND THE WILD AS BIOCAPITAL RESERVE

To understand what was now happening, we need to delve into the white paper and the situation from which it was speaking (Skinner 1969) in even more detail. For as we will see, there was severe trouble in this new economy, leading to massive death and loss in exactly what one had now invested so heavily in, the juvenile fish. The care for the living organisms that the Member of Parliament, Frøiland, had pointed to as so special in this new economy, was seriously failing. Most problematic were the conditions of the so-called *settefisk*, juvenile fish hatched and reared for the purposes of fish farming. These were produced at one locality, then cultured further and put into production elsewhere. They were vagabonds of sorts, on the move from one locality to the other, not by their own means or instincts, but by being transported around as an integral part of the production process. When large enough, they were released into net pens, where they were to remain until they reached their so-called slaughter weight. The problem, however, was that the juvenile fish were dying, in vast quantities. This happened while still staying in fresh water, and often during transport from one locality to the other. In fact, *seven out of ten* fish died at this stage (St. meld. 1986–1987, 20). And as production volumes were growing rapidly, so was the need for these juvenile, production-input fish. Soon, the amounts needed had vastly outgrown what national producers could provide. Consequently, the import of huge numbers of juveniles from abroad was allowed.

Up until 1983, half a million juvenile salmon fish (also called smolt) had been imported yearly. By 1984, this had risen to two million individuals (St. meld. 1986–1987, 21), in 1985 to 6.5 million individuals. While the ecosystem was established as a limit, the mass death of juvenile fish apparently was not. What turned out to be disastrous, it was noted, was that the imported fish were not always disease-free: In 1985, the disease furunculosis produced severe outbreaks at production facilities (St. meld. 1986–1987, 21), creating further sites and events of mass death among the farmed fish (see also Hovland 2014). Taken together, the whole aquaculture enterprise was, it was acknowledged, in a "critical phase" (St. meld. 1986–1987, 23).

One of the many problems that followed was an extensive use of medication. Unless a reduction were achieved, the white paper warned, restricting production might be the policy result (St. meld. 1986–1987, 24–25). The water environment was emerging as a large and growing problem caused by several factors. First, the growing aquaculture industry was polluting the water with its fish medication and, second, with the feed being given to the fish. Only part of the feed being put into the net pens contributed directly to the growth of the fish, while a significant part was added to the "recipient" environment adjacent to the fish farming facilities (St. meld. 1986–1987, 25). Third, dead and disease-riddled fish, often in huge amounts, were a problem, and regulations on "how such waste should be handled" were lacking (St. meld. 1986–1987, 25). As stated by the white paper, "many of the operating facilities" were "set back by environmental and disease problems" (St. meld. 1986–1987, 50). Some areas were worse off than others, and areas with a high density of fish and net pens were "showing signs of overload" (St. meld. 1986–1987, 50). In some places, production was now decreasing due to facilities being situated at "unfavorable localities" (St. meld. 1986–1987, 50).

The problems were directly linked to what had become established policy. In the very year that the 1985 NOU was published, a permanent aquaculture law was passed in Parliament, its purpose being to enable the government of a more modest and controlled growth. A whole set of issues had to be subjected to regulation and improvement. For instance, it became imperative to regulate the number of fish allowed in each net pen. This was to improve the environment for the fish, increasing their freedom of movement and reducing stress, to prevent the spread of pollution, and avoid the outbreak of disease. Simultaneously, however, the law represented a liberation, as it opened for a sharp increase in the production of juvenile fish. If the criteria regarding disease and pollution were in order, the law established, license *was* to be granted (St. meld. 1986–1987, 21; see also Kolle 2014, 175). In response, and mainly due to the high revenues to be had, the production capacity was rapidly increased.

The 1987 white paper was overflowing with environmental trouble and problems. nevertheless, what came out abundantly clear, was that on the economy side, small was no longer beautiful (cf. Schumacher 1973). The value ordering of the preceding decade, with its demands upon the industry to stay local and be owned locally, was now deemed more of a hindrance than a valued justification to pursuing aquaculture. One of the key questions, and "increasingly more relevant" (St. meld. 1986–1987, 30), was rather to pursue the advantages of a so-called economy of scale. Demand for capital was large, it was stated, and aquaculture was a "capital-intensive industry" (St. meld. 1986–1987, 30). The capital that was seen to be lacking was notably not investors' money, but the availability of exactly the type of biocapital we discussed in chapter 3—juvenile fish: "Access to seaworthy smolt has in recent years been the most severe bottleneck within aquaculture" (St. meld. 1986–1987, 20). On the financial side, the situation was bright, and radically altered from what had earlier been the case, when private banks had been "reluctant," as it was noted, toward the aquaculture industry (St. meld. 1986–1987, 31). During the 1980s, the banks had instead come to compete for fish farmers as their customers. Equity capital had likewise ceased to be a problem, as investors had been "lining up" to become partners in license holding companies (St. meld. 1986–1987, 32). On top of this came public money—the most important source of capital for the new aquaculture business—with the state providing loans, investment grants, and loan guarantees (St. meld. 1986–1987, 30). Revenues were high, especially at the juvenile-producing facilities, but also at the facilities producing the end product, the so-called food fish.

As revenues grew, the fish farming industry was beginning to exert pressure to repeal the regulatory licensing regime that aquaculture was based on (St. meld. 1986–1987, 34). The value ordering of intimate entanglements between local investors, managers, and owners was loosening up.

In place of an orientation toward staying local in the meaning of funding, management, and ownership of the aquaculture enterprise, there was now a push toward a rather different locally ordered system aimed at a different objective—namely, that of decentralizing the authority to *grant* facility licenses (St. meld. 1986–1987, 35). Instead of relying on expert judgments that were based centrally, it was reasoned, it was better to let decisions be made locally. Simultaneously, the size of the production facilities was growing, in 1987 being close to three times larger than only a few years back (St. meld. 1986–1987, 35).

Again, it is as if the white paper is in tension with itself. On the one hand, it reasons along the lines of the modesty ordering that this chapter describes: the aquaculture enterprise must and shall continue to be based locally. The whole enterprise is justified on this premise, that of building local communities. This purpose is also what justifies the flow of public capital from the state to the industry, most notably by way of the State's Industry and District Development Fund (Distriktenes Utbyggingsfond), a public agency that had as one of its primary tasks was to bolster the economies outside the major urban centers (see also chapter 5). On the other hand, the white paper carves out a space for change, a change that seems well underway already, and so the white paper also prepares to loosen the link between aquaculture and the small and local, and instead to move toward going big and growing large and, to that end, growing more liberal—for instance, with respect to the granting of concessions. It is toward the end of encountering this economization of the ocean that the eco-system vocabulary comes in as a form of counter-valuation.

Also with regard to the issue of biocapital—the fish biology on which the whole business was relying—it is as if two different authors or authorities have been writing the white paper simultaneously (St. meld. 1986–1987, 28–29). On the one hand, there is a modesty ordering, on the other hand, a move for speed and what can be gained by, for instance, breeding. Still, the white paper is somewhat more sober in its approach than what we observed in the blue book and the ocean growing approach we examined above. The wild species here are more than simply a material to be enhanced and modified. They also seem to have value apart from that of being domesticated and farmed, and even act as a form of capital, as it were, *in reserve*.

The breeding of today's well-adapted animals in agriculture has taken many hundreds of years, the white paper reasons, and so wild fish species represent “irreplaceable gene reservoirs” (St. meld. 1986–1987, 28); banks from which important traits can be retrieved and, eventually, “crossed” into the cultured version of the species. A strong and versatile ocean growing industry, depends not only on the mapping of these gene reservoirs, but on the preservation of them (St. meld. 1986–1987, 28). The reason for this is stated straightforwardly. The moment when a fish species is “brought into culture,” the white paper states, “the basis for a great deal of the natural selection disappears” (St. meld. 1986–1987, 28). This is simply because competition over food, breeding grounds, and mating is being eradicated for the benefit of selection “based on economically important traits.” This procedure, it is remarked, is the very foundation of the already started “genetic value enhancement” work in the early 1970s. Despite the trouble this might cause, the enhancement strategy is simply taken as a given. The search for profitability and cost reduction leads to strategies for enhancing value in the direction of developing a fish that, it is noted, “can exploit feed more efficiently” (St. meld. 1986–1987, 28–29). This is precisely what we refer to as co-modification: commodification as a relational process in which the species in question is carefully modified in intimate exchanges with economic considerations and market demands. In this way, co-modification is made integral to the process of economization. And it is made to depend, we learn, on enhancing the value of the species, for such economization ends.

A NEW VALUE ORDERING?

It is tempting to tell the above story in the form of a market logic that forcefully economizes and subjects nature to a path where it is near-predetermined to become capital and to be enrolled in a larger endeavor, making all things biological into capital. And surely, this is part of the story. It is however useful to dwell with what we have been observing, and what more there is to these events and our narrative. We started out this chapter with what we referred to as a scholarly intervention: not taking for granted, at the outset, that nature was already enrolled in the economy. Inspired by Schabas (2005) and Foucault ([1970] 1974), who in their different ways have analyzed how nature, biology, and economy developed into distinct fields and objects of knowledge, we wanted to show how a struggle to work on these fields and possibly bring them together can take place. Such co-modifications, we argue, happen at a myriad of sites: inside science, in experimental ponds, deep inside fjords, buried in business strategies, and—as this chapter has sought to show—in documents that, again in a myriad of ways, get linked up with a political machinery. These are sites not only for the ordering and reordering of social realities but for the ordering of nature and the biological, the species that without such interventions would simply continue to move by their own instincts. Policy documents and other documents close to the political apparatus, we show, are sites where such interventions are made to happen and where we can "go" to examine the work of co-modification. Simultaneously, policy documents are tools—tools that order and shape realities beyond themselves, and that also *value*, direct, and assess which moves to take and what a good order or "a good economy" might be.

Opening up policy documents to scrutiny—not only in a chase for the motivations, interests, and positions that are assumed to lie behind them, or treating their contents as either simply information or, the opposite, the expression of a certain ideology or pre-inscribed discourse—is vital in order to acknowledge how large transformations are the results of a series of wordy and worldly struggles and actions that could have been done otherwise. In related ways to how we observed in chapter 3 that various ocean sites—like ponds and net pens—can work as sites for the making of biocapital, policy documents can be sites for the manufacturing of biocapital too. In our case, we observe how policy documents work as sites where "the bio" is worked upon from the angle of having potential as capital and for a future accumulation of capital. The value ordering, then, is about enhancing the value of the species by bringing it into culture, though not *any* culture, but subjecting it, as we saw, to *commercial* culture.

It is sometimes argued that parliaments do not do politics of nature, at least not of any "real" or particularly interesting sort (Latour 2007), indicating that policy sites stay within their social realm and procedure, or simply rely on facts provided by the natural sciences. What we can observe throughout our reading of the NOU reports, the blue book, and the ensuing white paper is something quite different. Rather than being simply about the social order, or about the ordering of what is taken for granted as natural facts, they are sites that enact specific versions of nature and that order and reorder these (see Asdal 2015b; Asdal and Hobæk 2016). They are also sites of co-modification—sites where nature and the economic are drawn together and made to co-modify one another. For instance, the interventions in the blue book toward breeding were yet another way of conjuring nature realities that work in tandem with ways of conjuring economic realities.

Importantly, our analysis does not stop at considering policy documents as sites for the value ordering of nature. We also suggest that nature, rather than having a towering presence from the outset, was carefully manufactured and inserted into the policy circuitry of documents, as a thing to consider, account for, and be aware of. We have showed how this happened in a multitude of ways, including that of enacting nature as an entity with a given "carrying capacity," as a "production system," and as an "ecological system." The ocean was moreover articulated as a site of physical barriers, similar to those to be found on land. It could perhaps appear, it was reasoned, as if "the ocean was without limits." "Barriers" did exist, however, "invisible in the form of huge depths, temperature zones, uneven distribution of feed, etc." In other words, the ocean not only was made up of economic zones; there were what we can think of as nature zones as well. Even if these were less visible to the human eye, they were physical barriers to the species living in the ocean: "For most marine animal and plant-species these invisible borders are impossible to transgress." Moreover, a certain balance between species already existed, and, it was noted, "the balance in such a system" could "relatively easily be destroyed" (St. meld. 1986–1987, 69).

Interestingly, both the 1985 NOU report and the white paper submitted to Parliament two years later included a glossary. The glossaries were slightly, yet noticeably, different. The glossary of the NOU consisted mainly of words that were linked to aquaculture, acting as a kind of manual to this new economy and culture (NOU 1985, 7–8). The glossary of the white paper also consisted mainly of words from the world of aquaculture, but a few more words were added: "genetic variation," "stress," "carrying capacity," "recipient," and then, perhaps most importantly, "ecosystem" (St. meld. 1986–1987, 16–17). It is as if the white paper says, "Look, there is an ocean nature here, it has an ecology, its organisms can be stressed, it has a certain carrying capacity (it cannot take it all), it can be polluted, and it is actually quite important that there is a certain genetic variation to what we have here." In fact, as for the latter, it says almost exactly this, in its own straightforward and factual way: "The genetic variations within and between the populations are decisive for the species' ability to adapt to environmental variations in space and time" (St. meld. 1986–1987, 16). This also means that when landing in Parliament, the white paper takes on the distinct quality of being not only a site for the ordering of nature and for the co-modification of nature and economy. It is also a tool, an instruction manual for understanding how nature works, and how such a nature system might be vulnerable, become stressed, and have limits to what it can carry.

To be sure, it is easy not only to observe but also to critique how the above is not "nature" per se, but a specific version of it; a version represented through the science vocabulary of biology and ecology. Notions such as carrying capacity, biomass, and stress are not neutral. Nor do they spring directly and unmediated from nature "itself." Indeed, throughout this chapter we have shown examples of words that capture nature and the economy interchangeably. Yet, to defend nature's purity is not our point here. What we have instead aimed to show is how nature, in this specific version, emerges within political bodies and through the sites of policy documents. And moreover, how this distinct nature emerges as a relational effect of the emergence of a strongly interventionist process of economization. Limits to growth and interventions upon nature are made and enacted in the meeting, the encounter, with economy. Let's keep this in mind when we address, analyze, and seek to understand not only how nature works but also how knowledge and policy work: that nature and the economy are enacted in tandem with one another.

The turn toward ocean growing and, with this, a quite different value ordering, did not come easily. There were failure, trouble, disruption, loss, diseases, mass death, and major environmental problems. More than a new order, there was disorder; more than a logic, there were confusion and movements in quite radically different directions, simultaneously. Somewhere amid this disorder and trouble, we can nevertheless trace the contours of a new value ordering—and furthermore, a turn to what the next chapter will identify as an innovation economy, "innovation" becoming the overriding ambition and "buzz word" of ocean growing and for the domestication and farming of cod.

5 INTO INNOVATION

The history of the cod fisheries is a history of fluctuations, the ocean teaming with fish some years, "black" in its absence in others. What the fishers call the "white gold" of the ocean is not something that is mined predictably. The cod offers itself in the form of highs and lows, as it has done for centuries, which also spurred the late nineteenth-century ideas of taking control over cod stocks. As we saw in chapter 3, this included Halvor Dannevig's experiments in Flødevigen, but it was also expressed in research reports delivered to Parliament (Asdal and Hobæk 2016). Having spent days on end in an open boat, peering over the gunwale, and studying the precarious life of fertilized cod eggs, the zoologist Georg Ossian Sars raised the question of whether one "by art, should be auxiliary to Nature, to secure oneself for the future against other lean years" (1869, 18; authors' translation). Such lean years, he underlined, had a marked impact not only on those directly vested in the fisheries but on the prosperity of the entire nation. With this, Sars formulated a codependence between the riches of the sea and the wealth of the nation, a relationship that we explore in this chapter as a rationale for steering investment capital toward the aquaculture industry—capital, notably, that not only was to flow in the right directions and in sufficient quantities; it was also *to behave* in certain ways, be "competent," "patient," and "risk-willing." Only then, it seems, was capital able to draw cod biology and markets together and *into innovation*.

Today, the work of Sars, Dannevig, and other early zoologists and marine biologists is routinely cited in the introductory pages of aquaculture research reports, policy documents, and innovation strategies. Enacted as a first push toward or as a precursor to the Norwegian aquaculture industry, their work has come to constitute a form of nineteenth-century "origin story" (Haraway 1989, 1997) that, at least since the 1980s, has accompanied documents advocating for aquaculture research, development, and investment. In some cases, this purportedly "long history" of ambitions to cultivate the ocean works as a form of "fun fact" preamble to more technical considerations. In others, it is used to justify the idea of Norway as a aquaculture nation and, correspondingly, that the state should invest funds in its development. Around 2000, however, something interesting happens to the way of telling this nineteenth-century origin story, as a second type of origin story is then being activated. Often told alongside the one celebrating the "pioneer" activities of men like Sars and Dannevig, this second story is about the recent and unprecedented progress and success of farmed salmon. A model species for all farmed fish to come, it now seemed, the salmon had since the 1980s proved that the fish farming industry could grow fast and be profitable. As one proceeded with renewed efforts to farm the Atlantic cod, in 2000 and the following decade, the idea therefore was to make it more like a salmon. If helped along by the experiences and know-how of salmon farming, the cod was now believed to finally be ready to become part of the aquaculture industry. Another impetus for a renewed effort toward cod farming was that the fisheries had since the end of the 1980s been quite bad. Once again, fishers were talking about "black ocean," and policy makers about a resource crisis caused by overfishing. To remedy the situation the fishing quotas were cut significantly, and the old vision of reproducing the cod artificially was yet again on the agenda. Could the "slumbering giant" (NOU 1985, 12; see also chapter 4) be awakened, making cod farming the next big thing?

Scientists were optimistic and politicians followed suit. The Prime Minister at the time, Jens Stoltenberg, promised in 2001 nothing but a "revolution" in cod farming (Asdal 2015b; Enoksen 2017). The Food and Agriculture Organization of the United Nations (FAO) pointed to cod as the world's most interesting species to be cultivated (Enoksen 2017). Writing about cod farming in 2000, the business newspaper *Dagens Næringsliv* furthermore stated that cod was now the "hottest object of investment" (*Dagens Næringsliv*, October 28, 2000, cited in Enoksen 2017, 98), a statement that shows that its promise was recognized not only by policy makers but also by investors. And further, that the cod needed, precisely, investment. The establishment of cod farming was not only considered promising but also costly, and attracting capital was key to putting the enterprise in motion. The issue therefore was not only *marketization*, turning the cod into a viably market commodity, but also to make it an object of investment, a process we approach in this chapter as a form of *capitalization*. This double operation of marketization and capitalization, we suggest, is key to understanding the version of economization that this chapter examines—the innovation paradigm—and how it works by drawing nature into economy by way of co-modifications. Following the cod as efforts are once again being made to turn it into a profitable farmed fish, the overriding objective of the chapter is to simultaneously analyze how the cod is taken *into innovation* and to discern the version of economization that guides the innovation paradigm.

THE DOUBLE OPERATION OF CAPITALIZATION AND MARKETIZATION

The "immense continent" of capitalization has been sadly neglected in anthropological and sociological research, argue the authors of the book titled, exactly, *Capitalization* (Muniesa et al. 2017, 13). What they regret is that commodification or marketization has been employed as the crux of economic analysis at the expense of capitalization. The analysis of this chapter seeks to remedy some of this former neglect. Yet, to make an opposition between capitalization and marketization does not fully serve our end. Instead, what our analysis will show is that the double operation of marketization and capitalization, and the bringing-together of these, is a core quality of the innovation paradigm. This furthermore speaks to the notion of co-modification, which we employ throughout this book to examine how nature and the economy work on one another. An innovation economy, our analysis suggests, is about co-modification par excellence. It is its key mode of ordering. The chapter's analysis thereby follows the direction set by an earlier study that demonstrates how co-modifications can operate to bring the sites of production and markets together (Asdal and Cointe 2021). To understand the version of economization that we delineate as the innovation paradigm, we suggest, we need to grasp how this paradigm aims toward working on both "sides"—of making both capital *and* markets behave and grow in distinct ways. The chapter thereby identifies operations of capitalization, yet also intense operations of market work, and discerns how they, together, are entangled in various co-modifications.

Empirically and analytically, the chapter moves in three steps:

First, it follows innovation as it is framed as a new form of economization at the turn of the millennium. Importantly, this move delineates our object of study as not an innovation economy in and of itself, but an innovation economy that emerges as a form of public policy. In line with the overall analysis of the book, we show how versions of economization are often intimately coupled with the state machinery. In the case of innovation, the result of this is the emergence of quite complex and hybrid relations across private and public actors and agencies (see also chapters 6 and 7). This speaks to how the state is asked to support innovation, but also, and just as significantly, to how the state is worked upon as part of the innovation paradigm. The state, we show, is challenged to take it upon itself to act as an investor and provide the right form of capital, as well as to provide tools that can enable collective entrepreneurial action across the domains of the public and the private.

Second, we follow the double and indeed quite ambiguous life of capital. As we describe above, cod farming was considered as both an investment opportunity and as capital-demanding. On the one hand, capital is put forward as that which is to *fund* innovation. On the other, it is also seen to represent the very object that monetary capital is seeking to invest in—a cod that is also, potentially, a form of biocapital. So, what does it mean for the cod to become a form of capital? In chapter 3 we began to examine this by considering the rearing and raising of cod as stocks of biocapital. We stayed close to the cod and carefully described both how the cod was made to behave and how the cod resisted and actively took part in efforts to turn it into a farm animal. This chapter extends this attention toward agency and observes how it is made to encapsulate the monetary form of capital too. This is not to say that capital is being equipped with human will or intentions. We simply show how capital is rendered as having specific capacities and qualities. In this, we follow the Greimasian semiotics (Greimas and Courtés 1982) that served as a foundation to actor-network theory and its demonstrations of how a multitude of "actants," human as well as nonhuman, can make a difference and enable something to happen (Asdal and Jordheim 2018; Law and Hassard 1999). In related ways we observe how capital is rendered active and endowed with new and quite specific qualities and capacities. We argue that such shifts in capital capacities serve as a condition of possibility for what we in this chapter identify as an innovation economy.

Third, we follow how the Atlantic cod is subjected to innovation and acted upon by way of a document species that we name "cod plans." Together with interviews with investors, cod farmers, and public agencies, these cod plans (RCN/SND 2001, 2003; RCN/IN 2006; RCN/IN/NSRF 2009) make up the empirical material of this chapter and are analyzed as both sites and tools of innovation (Asdal 2015). The first cod plan, issued in 2001, was published in cooperation with the state agency called the State's Industry and District Development Fund (SND) and the Research Council of Norway (RCN). From 2006, the SND was replaced by the new public agency "Innovation Norway," which is also reflected in that Innovation Norway becomes an issuer of the 2006 and 2009 cod plans. These latter two plans came into being through a tight coupling with a public institutional reordering that was turning "innovation" into public policy (see also Gulbrandsen 2011; Teigen 2019). We analyze these plans and strategies as being sites of action, yet also how they work as what we examine in this book as tools of valuation—tools that actively register, consider, move, praise, direct, judge, and qualify the various elements that are drawn into the process of making this specific innovation, cod farming, happen.

INNOVATION ECONOMY AS PUBLIC POLICY

Two events took place at the turn of the millennium that signaled an innovation economy in the making. First, the innovation paradigm was institutionalized, when in 2004 Innovation Norway replaced the State's Industry and District Development Fund. This implied that "the districts," which in Norway points to communities outside the main urban hubs, were no longer the main targets for economic development. Irrespective of where, Innovation Norway was to promote innovation on a national scale and support a stronger export orientation (Teigen 2019). Second, the innovation paradigm presented itself with much confidence at a policy level: It is not often that white papers or propositions to the Norwegian Parliament take it upon themselves to educate their audience on new economic theories. Yet this is precisely what the proposition put forward by the Ministry of Business and Commerce in spring 2003 did (St. prp. 2002–2003; Teigen

2019). Titled *Measures for an Innovative and Inventive Business Sector,* the proposition argues quite bluntly for innovation and innovation theory as the new way of reasoning in economy policy. Building in part on the earlier Official Norwegian Report (NOU) from 2000, *A New Go for Value Creation* (NOU 2000), with a mission toward creating "a better climate for private investment," the proposition to Parliament puts innovation at the center stage of the policy agenda.

An innovation, the proposition states, is something new—a product, a production process, or a new organizational form—that is either launched in markets or put to work in production to create economic values (St. prp. 2002–2003, 7–10). The theory of innovation, it is further stated, starts from the understanding that innovation is a key component in value creation. But how to make such innovations to happen? According to the proposition, a precondition for innovation is "interaction," here taken to include competition as well as cooperation and to encompass not only interaction between private actors but also between private and public actors (St. prp. 2002–2003, 7). That of generating research is put forward as a key objective, and so is the importance of being "dynamic." The latter may perhaps appear as somewhat unclear, but being dynamic is ascribed both a very practical significance and a theoretical one, and is key to challenging neoclassical economics as the then dominant paradigm in economic reasoning.

The theoretical importance of being dynamic is explained vis-à-vis a core issue in economics: situations when markets fail to work efficiently. This is what economic theory calls "market failure," which is also how it is described in the proposition. The critique and problem, as understood from the side of innovation theory, is that neoclassical economics approaches market failure as a rather static condition. Innovation theories do not see the world in such static terms and emphasize instead a more dynamic approach, the proposition informs. Innovation happens in a dynamic interplay and in between many different actors (St. prp. 2002–2003, 15). Included in this dynamism, the proposition underlines, is the risk of failing. It was important to be aware, therefore, that enterprises that try out new ideas, but fail, nevertheless contribute to increased innovation in the business sector. Entrepreneurs that had failed would, for instance, often acquire valuable competence that could be of use in later projects (St. prp. 2002–2003, 21). Not only in theory, but also in practice is being dynamic of key importance, which is also reflected in how the proposition to Parliament put forward the innovation paradigm as a bundle of guidelines and imperatives for action. The proposition points to several key measures: Public authorities must create conditions for efficient competition and stimulate what is called "co-creation and learning." Cooperation between actors is deemed essential and can be encouraged by coupling young businesses, entrepreneurs, and research environments with actors with market competence, market access, and so-called competent capital (St. prp. 2002–2003, 21). The challenge, then, is not only to attract capital but to attract capital with the *right* qualities, a capital that will *behave,* and in a good manner. Such qualified capital, however, is lacking. It seems, it is noted, that not only do "growth companies" (*vekstforetak*) lack capital that is "competent," they also lack capital that is "patient" and "risk-willing." More generally, growth companies are understood to be "underfinanced"—meaning that projects that ought to have been realized are not being funded (St. prp. 2002–2003, 13).

THE NEW WORTH OF CAPITAL

In the scholarly literature on innovation, it has been emphasized that innovation is an expensive process (O'Sullivan 2005). Significant amounts of capital resources must therefore be drawn toward initiating, directing, and sustaining the innovation process. The significance of capital quantities is underlined also from a very different scholarly end. In *The Enigma of Capital and the Crises of Capitalism,* David Harvey describes capital as "the lifeblood that flows through the body politic of all those societies we call capitalist, spreading out, sometimes as a trickle and other times as a flood, into every nook and cranny of the inhabited world" (2010, vi). Independent of whether capital is understood as something scarce that needs to be attracted, or as flow and even overflow in need of somewhere to invest, capital is commonly presented as something that exists, precisely, in *quantities,* not with *qualities* of its own. Yet the description of capital as preferably being "competent," as we saw above, signals something different. Capital also must take on certain qualities, this tells us, and specific qualities are deemed necessary to accomplish innovation. In fact, it is as if investment capital, be it public or private, is taking on a whole new set of qualities. That of *qualifying* capital is made integral to innovation as a version of economization. Through such qualifying procedures, the worth of capital is changing—a change that is a vital feature of the innovation economy. As we will see from the ensuing analysis, especially public capital is in the Norwegian context qualified as good capital with the right qualities to assist and enable innovation.

By this we do not take "worth" to describe the changing value of money through, for instance, inflation or changing currency rates. What we are after are qualitative considerations of capital. In the 1970s and early 1980s, at the start of the modern aquaculture enterprise, capital, or a specific version of it, was considered a definite evil. In fact, the aim was to keep it away from the aquaculture enterprise (Kolle 2014, 153). Expressions like "foreign capital," "capital interests," and "big capital" signaled how capital was in tension with and opposition to what was considered of worth in aquaculture and, further, how it was not deemed part of the good economy (Asdal et al. 2021). Big capital or large capital interests were predominantly a threat and a problem, and as if in direct conflict with real worth and real value in aquaculture. This was altered in the decade and version of economization that this chapter examines. By the turn of the millennium and with the advent of the innovation paradigm, the aquaculture enterprise had come to be considered as a "capital-demanding" and "capital-intensive" activity, an enterprise in *need* of capital, and not only any capital, but capital that was "risk willing"—capital that was directed toward taking risks and, implicitly, capital that could be lost. Capital was furthermore something to be secured for the industry, an industry that had ambitions and sometimes even the ability to demonstrate what was called "financial muscles." Big capital had, in other words, turned good. Both its quantity and its quality were now a condition of possibility for the whole aquaculture enterprise. Capital was shifting character and quality, becoming a definite good, but also a thing that could be improved, to become, as we saw above, qualified, or "competent." This tells us that to consider the flows of capital is not sufficient. We must also consider the changing qualities and the worth of capital, and particularly so if we are to grasp innovation as a version of economization.

The book *Capitalization* (Muniesa et al. 2017; see also Chiapello 2020) can help us move the analysis in this direction. Parallel to our reasoning also in chapters 3 and 4, it argues that capitalization involves moving something in a particular direction, *into* the economy. What we suggest here is to consider, in addition to this, the qualitative properties of investment capital and that of instilling in capital new value and the right and good qualities. In analyzing such operations, we now turn to consider the above-mentioned series of cod plans. Issued in 2001, 2003, 2006, and 2009, these work as tools of valuation that seek to entice and steer innovation in cod farming, enabling its transition from a start-up industry to large-scale production. However, as the ensuing analysis will show, the cod plans are not only involved in operations of capitalization, but profoundly also acting toward marketization. This double operation we argue, is key to innovation as a version of economization.

ACTING FOR FARMED COD: INNOVATION OPERATIONS

Most document species—disregarding their sometimes ascetic, dry, and quite restrictive form—actively work upon the issues they raise (Asdal 2015a). Yet the documents of this chapter, the cod plans, are as if *particularly* action-oriented. As we will show in further detail below, their very mode of operation is *to do* economy, to act and to make act and this, we will argue, is one of the ways in which they are part and parcel of an innovation economy and paradigm: They are intensely performative and action-oriented.

Titled *Farming of Cod: Strategy for Coordinated Effort*, the first of the cod plans instantly stages the cod as the object to be worked on: on the cover page, it is captured as swimming toward the reader (see figure 5.1). The cod is not presented au naturel, however. Its forehead is stamped with the brand of the Norwegian Seafood Council,[1] a state-owned agency working to enhance the export value of Norwegian seafood products (for further discussion of the Norwegian Seafood Council, see chapters 6 and 7). The text of the brand, "Norge—Seafood from Norway," reflects a long-standing, international strategy of marketing Norwegian seafood as having superior quality. The brand can as such be seen as a judgment device, that is, a device that "helps" consumers make choices by offering information and knowledge (Karpik 2010, 44). Still, the image of the branded cod is here not directed toward the consumer, at least not directly. More than aiding or steering consumer judgment, this is about valuing the farmed cod for its potential to become part of the Norwegian export economy. Imprinted on the cod and the cod plan, the Seafood Council's brand acts as a tool of valuation, validating the cod as a particularly promising species and object of innovation. As claimed by the cod plan, "There exists a large international market for cod, and a wide range of cod-related products are in demand in

FIGURE 5.1
Image of the cod from the front page of the cod plan *Farming of Cod: Strategy for Coordinated Effort by SND and the Research Council of Norway, 2010–2010* (RCN/SND 2001, front matter).

large parts of the world" (RCN/SND 2001, 2). In twenty years, the cod plan estimates, cod farming can produce values that equal those of the current salmon production (RCN/SND 2001, 2). In other words, the cod has the potential to be transformed into a salmon. This future, we are made to understand, is made possible by a rich, proud, and favorable environment, and so the cod's history, its surroundings and then also the origin story to which we referred above, is brought into the innovation operation:

> Cod farming is an industry where Norway, with its geography, infrastructure and competence in fish farming will have huge competitive advantages. Norwegian research and development environments have a solid competence in cod as a farmed species. Cod is probably the fish species that can most quickly reach large production volumes in farming because modified salmon technology can be used in the phase of growth in the sea. Today there is huge interest amongst several commercial actors to commence large-scale cod farming, and there is a lot of private capital available for investments in production units. The probability of succeeding in cod farming is significantly larger today than it was 10–15 years ago. (RCN/SND 2001, 2)

The capital is available, the cod is suitable, and—just as in Michel Callon's (1984) classic story of the domestication of scallops—there exists a market out there, inhabited by hungry consumers eagerly awaiting a much-appreciated commodity. The earlier worries that the production capacity could exceed what the market could absorb are now quite clearly gone. The timing is advantageous and better than before, and there are huge competitive advantages to be had for Norway too. As is also quite strongly suggested by the title of the cod plan, *Farming of Cod: Strategy for Coordinated Effort*, this is a version of economization that rests on bringing together and fostering cooperation. Science, technology, competence, entrepreneurship, investment capital, even natural conditions and geography, are to be aligned toward the aim of innovation. Emphasizing coordination and cooperation, the plan works nicely in line with the innovation economy paradigm we observed above, at the policy level.

What the proposition to the Parliament did not consider, but which is made very clear in the cod plan, is that the list of cooperating agents of innovation was lacking one key agent, namely the material or nature to work from. And surely, to make the cod cooperate is one of the main tasks that the cod plans set out to solve. Toward this end, the 2001 cod plan and the ones to follow bring forth a series of problems to be solved. There are problems tied to the production side of cod farming. The capacity of cod fry producers is too low for large-scale industrialization; there are problems with illness and mass death in the net pens; the cod are escaping in huge numbers; or they are being eaten by predators, sometimes even by each other. The relatively few cod that make it to slaughter are often of poor quality, their liver is too large, and their flesh too soft. Despite the great optimism held on behalf of the cod, the great difficulty of turning what here appears to be a not-so-domesticated fish into a species suitable for industrial farming purposes is acknowledged.

To make cod farming profitable, the cod is fleshed out as in need of modifications that can turn it into a successful innovation. Cod fry need to be grown on a large scale and at a sufficiently low price. Breeding, or so-called selection for increased productivity needs to be systematized to enhance the cod's capacities. The cod's eating habits must be altered, for instance the abovementioned cannibalism, but also the amount of feed the cod requires per kilogram of growth. Diseases and parasites need to be better controlled and the cod itself must be developed into a more resistant organism. Finally, there is the issue of early onset of sexual maturation among the cod in captivity, which is described as "probably the most important bottleneck in achieving good economy in fish farming" (RCN/SND 2001, 15). Counteracting early sexual maturation is also one of the key tasks identified by the cod plan, on behalf of the "coordinated effort" it seeks to entice. Clearly, this is not only about working on markets but about investing in cod and in turning it into a form of capital, giving it the capacity to behave so that profitability can be achieved. As such this follows up on chapter 3, where we described how the domesticated cod is constituted as a form of biocapital. What we observe in the 2001 cod plan, as well as in the ones that follow it, is a next step in the co-modification operation. Here, flows of investment capital—capital whose desired qualities are abundance, patience, competence, and risk-willingness—are set up to meet the farmed cod in ways that can help realize its promissory qualities. It is the co-modification of cod qualities with a capital that is willing to take the risk, to try and retry, and to act on and stay with the cod that is what can stabilize the farmed cod as a commodity. In this way, this specific kind of co-modification stands out as a key condition to turning the farming of cod into an economy.

TIMING TROUBLE AND ACTING DYNAMICALLY

The first cod plan from 2001 does not so much explicate why sexual maturation is a problem, but points to the necessity of continuing already initiated research to solve it. In this research, the problem is delineated by contrasting the farmed version of the cod with the wild Atlantic cod (Asdal 2015b; Taranger et al. 2010, 487; see also Taranger et al. 2006 for a later publication on this issue). The spawning period for most stocks of Atlantic cod, it explains, is between January and April, but the cod's growth rate and puberty age vary between stocks. Sexual maturation is also influenced by the availability of prey and the temperature in the cod's environment. For instance, there is a difference between the northeast Arctic cod stock—the *skrei*—that usually spawns at an age between four and eight years, and the Norwegian coastal cod stocks, which spawn at three years and older. Domestication, however, not only eliminates these differences, but profoundly changes and accelerates the spawning time of the cod.

In the net pens, under what scientists defined as otherwise "normal growing conditions," the spawning starts at age two, sometimes even earlier, which impacts the productivity of cod farming negatively in several ways (Norberg 2002 cited in Asdal 2015b). The production of eggs in the female fish and of milt in the male are both highly energy-demanding. The cod's entrance into puberty is therefore associated with significant loss of body mass and deteriorating health, which increases its susceptibility to diseases and other spawning-associated causes of mortality. The weight loss also means that the harvest quality is significantly reduced, due to the loss of biomass and the decreased quality of the remaining meat. The so-called feed factor, which signifies how many kilograms of feed the cod needs per kilogram of growth, is raised and, alongside it, the cost per kilogram of cod produced. Finally, and crucially, this all happens before the cod has reached "market size," meaning that the growth of the fish stops before it has reached a size that makes it a viable market product (Norberg 2002 cited in Asdal 2015b, 179). Put bluntly, when the cod spawns, profits drop.

Much in line with the prospects for fish farming described in chapter 4, the cod is here valued as a modifiable material. The research on early sexual maturation among farmed cod is set to unveil the "mechanisms" of reproductive timing and, through this, identify ways of controlling these. Yet, and as signified by the researchers' concern with costs and, perhaps most prominently, by the notion of "market size," the problem targeted is not only the biological performance of the farmed cod. In a very concrete way, the two concerns of modifying cod biology and doing market work are targeted together. The question becomes how fast and at what cost a marketable size can be reached—and the answer is that the "onset of puberty must be delayed by at least one year" (Norberg 2002 cited in Asdal 2015b, 179).

A key aspect to innovation—its mode of operation—is co-modification; careful modifications that work simultaneously on nature, markets, and economy. In the case of the cod farming enterprise this is, on the one hand, about modifying markets in coordination with the cod biology. On the other hand, this is about modifying the flow and behavior of capital in intimate relations with the cod, which is itself turned into a form of biocapital. In both instances co-modification concerns how nature is taken into innovation as "modifiable material." Examining the cod plans we can observe how the different measures called for to modify the cod's biological performances are not simply about lowering production costs. They also aim to create a fish that can, ultimately, achieve higher prices. And surely, the cod plans do not stop at modifying the cod toward market demands. They also address how profitability in a future cod farming industry depends on working on markets so that they can be aligned with the affordances of the cod. On this point, however, the 2001 cod plan is somewhat less specific than the plans to follow.

A striking side to the cod plans as they were published, first in 2001, then in 2003, 2006, and finally in 2009, is how they are quite similar to one another. In fact, the 2003 cod plan can be read as close to a copy of the plan from 2001, and the 2006 plan as close to the 2003 plan. The publishing years change, but the title, contents, and problems addressed stay closely related, forming together the genre of innovation operations. This implies that more than simply repetitions, the plans are operations of coordination that take part in producing stability and keeping things together. The plans are tools of valuation that maintain the status of cod farming as a desirable project as they seek to spur innovative action toward its realization. Yet, reading the cod plans closely, we also see how they work toward this end by changing, if only a little, from the one plan to the next. This is modifying work that matters, as it quite carefully shifts, adjusts, and reorders as the plans take in the (few) successes and the (many) unforeseen problems that arise in the market, as well as the efforts taking place elsewhere to modify the cod. And clearly, there are problems that prove difficult to solve. Notably, the 2003 and 2006 cod plans nuance the blatant optimism of the 2001 plan by addressing more closely the challenges facing the cod farming industry. Indeed—and different from how the 2001 cod plan depicted possible opportunities and challenges—the 2003 and 2006 cod plans draw on experiences gained by actors engaged in both research and entrepreneurial activities. Through this, a certain feedback dynamism is created, where concerns voiced by the industry are absorbed as part of that which needs to be modified to facilitate further innovation. And indeed, the plans' way of operating is precisely that of acting upon problems as they are constantly emerging. This procedure of constant coordination and adjustment is also what is taken on board with regard to the co-modification of market work with market research.

CO-MODIFYING MARKET WORK WITH MARKET RESEARCH

At the outset, the market situation was, as we saw above, envisioned as good. The 2001 cod plan assumed that there was an already existing market inhabited by consumers demanding cod products (RCN/SND 2001). For instance, cod was described as having been an important seafood product for centuries, treasured by Norwegians for thousands of years for its quality. It was also pointed out that Norway was the leader in the production of farmed cod and a substantial increase in production was expected. Whereas the production of fry had been 500,000 in the year 2000, this was expected to rise to two million in 2002 and estimated to be at ten million in 2004 (RCN/SND 2003, 8). While such assumptions keep being referred to, in later cod plans the market enters the innovation strategies in a more detailed way, as does market research, which is increasingly underlined as an important task and challenge.

In the 2003 cod plan, it is optimistically noted how, through "market work and product development" it would be possible to reach higher prices for farmed fish than for ocean-captured cod (RCN/SND 2003, 12). To achieve this, so-called market-based product development ought to be intensified and the products developed in line with the needs and preferences in the market. However, as market research is taken into the 2003 cod plan, the optimistic scenarios are now delivered alongside notions of possible limitations and trouble. For instance, both chefs and consumers expected prices of farmed products to be considerably lower than those of "wild products" (RCN/SND 2003, 16). Responding to this issue, the 2003 cod plan casts these attitudes as something one needed to work on and change. Following up on this, the ensuing 2006 cod plan notes that to achieve "satisfactory" cod value, the consumers' willingness to pay for the farmed fish "had to be developed" (RCN/IN 2006, 29). So, how could the consumer be attracted to the farmed rather than the "wild" products? In seeking to answer these questions, the cod plans turned to market research and the insights it could offer into identifying and solving such market(ing) problems. For instance, expert panels set up to determine the sensory qualities of the farmed and ocean-captured versions of cod suggested that a longer starvation period before slaughter could make the sensory qualities of the farmed cod meat be more like that of the wild cod (Luten et al. 2002, 44). In other words, not only was it an option to modify the farmed cod to achieve an optimal *size*. Another option was to enhance its value in the market by increasing its *sensory* value, hence its praise value. However, the longer starvation period was not deemed sufficient (Luten et al. 2002, 44). Instead, other research considered other measures. Providing product information about the origin of the cod could perhaps improve how it was valued by consumers, as could perhaps that of providing the name of the catching vessel, farming location, and processing location (Luten et al. 2002, 59–60).

Surely, the cod plans are not only working *on behalf* of cod farming actors. Additional demands *on* them and the industry are also made. For whereas the 2001 cod plan engaged primarily with ways of modifying cod that could bring production costs down and market values up, something different is added to the 2003 and 2006 plans. Here, "environmental and ethical challenges" are included in the concerns that cod farmers must deal with. Additionally, "suitable ocean areas" are being portrayed as a "scarce commodity" (RCN/SND 2003, 17)—the ocean is becoming crowded, as we have noted in earlier chapters. Concerns are also raised about the farmed cod being a danger to the wild cod stocks. One issue is escapees, who can get mixed up with and reduce the genetic quality of wild stocks, or even infect them with diseases from the net pens. Another issue is premature spawning in the net pens, which means that fertilized eggs from the domesticated cod can drift into adjacent ocean environments. All sorts of boundaries are seemingly being crossed, putting nature "in the wild" at risk. Another problem that turns up is the welfare of the farmed cod. Contrary to the 2001 cod plan, which dealt only with its "sickness and health," the 2003 and 2006 plans also ask involved actors to consider and improve the "welfare" of the cod. However, and much in line with how Hobæk (2023) describes the status of ethical issues in the aquaculture industry, the concerns over animal welfare and environmental impact are not so much put forward as issues in and of themselves. They enter most strongly by way of the consumer and, more specifically, through consumer preferences identified by market research. Again, this can be seen as a form of co-modification operation. Market preferences are made to work on the cod's welfare and, by extension, on the conduct of the fish farmers: "Consumers are increasingly more conscious of health and environment, and the demands for safe food, documentation of product contents and traceability are continually being made more stringent" (RCN/SND 2003, 11). This way of bringing in the consumer is repeated in the 2006 plan, where the prognosis of "fast growth and a production of great significance" is also cited to call on the industry to begin developing "documentation that identifies cod as safe food" (RCN/IN 2006, 15). In this way, the notion of worth is extended; if this co-modification is to succeed, one must extend the register of worth to include how consumers are valuing more than simply the best price. Again, then, we see how the plans work as tools of valuation, enticing the cod farming industry to register what counts as valuable in the market and to include this in their future practices. Yet the consumer's preferences were not as welcoming as initially assumed, and the cod's appearance and affordances not so easily modifiable as one could have wanted. Could that of better co-ordination between the farmed cod and the wild cod, in markets, be a way out?

From the very beginning, the prospects of the farmed cod were formulated in relation to those of the cod in "the wild." For instance, that the situation in the early 2000s was deemed favorable for the start-up of cod farming was linked to the current state of cod stocks. Catches were low, fishing quotas were being reduced, and the volume of cod reaching markets were sinking. This situation, it was argued, paved the way for the farmed cod as a form of substitute. In the longer term, it might even make up a larger part of the production volume, "thus contributing to maintaining the delivery of raw material and the activities of the fish-processing industry along the coast and contributing to more stable deliveries to the markets" (RCN/IN 2006, 8; RCN/SND 2001, 5; 2003, 7). These aspirations also tie into the challenges associated with fresh cod being a seasonal commodity, and the perceived market opportunities of turning cod it into a year-round good. If both farmed and ocean captured cod could be marketed as "fresh cod," the cod plans envisioned, the farmed cod could moreover take on the more positive consumer perceptions of the "wild" cod. Rather than being made to stand out as a distinct and unique market product (cf. Karpik 2010), the strategy was to downplay the quality of being farmed and instead promote "fresh cod" from Norway. "Fresh," not "farmed," is hence made to stand out as the new commodity.

The cod in "the wild," however, were difficult to bring into this. In a manner of speaking, they were poorly coordinated with the innovation strategies. Instead, they began to return in larger numbers and were sold at what was considered as quite good market prices (RCN/IN/NSRF 2009, 11). Consequently, the farmed cod ceased to be offered as a substitute, but rather became *complementary* to the ocean-captured cod. As formulated in the 2009 cod plan: "Farmed cod and wild cod will complement each other and make it possible to offer European consumers fresh cod all year round. This is important for the fresh cod—both wild and farmed—to be given space on the supermarket shelves" (RCN/IN/NSRF 2009, 47). Moreover, the ocean-captured cod was in the 2009 cod plan no longer made part of "the past." On the contrary, the "wild" member of the cod family is presented as a valuable resource, a resource on which the farmed cod can capitalize: "The perception and judgment of the cod as a market product depend on the ability to support the farmed, as well as the wild, cod. They are both high-quality and sustainably governed products" (RCN/IN/NSRF 2009, 47). The move toward selling farmed cod as fresh consequently involved a double operation. On the one hand, it entailed a marketization whereby the cod was both presented as and actively modified into a commodity. On the other hand, it entailed capitalizing on the cod in the wild, adding value to the farmed cod by making it integral to the new entity—"fresh" cod from Norway. The strategy can be read as producing a neutralized cod or, expressed more actively, devaluing "the farmed" for the benefit of the "fresh" cod, making it a seamless part of an already highly valued environment, history, and commodity.

MOBILIZING AND ACTING WITH CAPITAL

Modifying markets to enable a successful meeting between the valuations of consumers and cod qualities, was not sufficient for cod farming to succeed. From the very start, attracting capital to fund it, and not only capital as such, but capital with the right qualities, was an essential part of enabling innovation. Cod farming was "capital-intensive" and a high-risk activity, the 2001 cod plan established (RCN/SND 2001, 2). What was in demand was capital that could enable the industry to go big—and what was in real demand was public funds (RCN/SND 2001, 2). "Success," the 2001 cod plan states, "will demand close and good communication and coordination between industry actors, researchers, and experts. Industry actors and researchers must together establish projects to solve production-related problem areas and bottlenecks" (RCN/SND 2001, 18). The quote is from the very last pages of the first cod plan, where the Research Council of Norway (RCN) and the State's Industry and District Development Fund (SND) delineate how they will prioritize and organize, and, importantly, what type of ventures they will fund: due to the high financial risk of entering cod farming, businesses must have a "sufficient capital base." Moreover, they should have the prerequisite of developing "large-scale production" (RCN/SND 2001, 19).

To be sure, there was a "carrot" at the end of the innovation "stick." The plans were meant not only to raise cod but to raise funds and to finance the modification of cod into a farmed species. In delineating what type of innovation scheme one would be willing to fund, the plans make a further move toward "coordination." To spur innovation, the aim is to enroll not only researchers and industry actors, but also the government and, notably, the two state agencies behind the plan. In the 2001 cod plan, this is conducted by describing cod farming as an early-stage, high-risk, capital-intensive industry in need of "risk discharge." Calculations are then provided to show how much capital is needed to "discharge" industry actors sufficiently (RCN/SND 2001, 10). By comparing the results of this calculation (NOK 334 million) to the funds allocated for the purpose (close to none), the plan identifies a "shortage of risk-willing public capital" (RCN/SND 2001, 10–11, 17), which, by the logics of the plan, it is up to the government to fill in its next state budget. The state is thereby called on as both a long-term provider of risk capital and an attractive investor, precisely due to its ability to provide the right form of capital: capital that is both patient and risk-willing, the latter entailing a form of funding that can be lost if the project fails. The 2001 cod plan then moves on to describe the role of the RCN and SND—here addressed as not only agencies for *allocating* such risk-willing capital but agencies that can act as a "driving force" in ensuring "perpetuity and gravity" in the public effort (RCN/SND 2001, 18). By taking on the task of promoting the cod farming industry to the government, and through this ensuring public innovation capital of the desired quality, the two agencies quite actively lend themselves to the idea that cod farming should be invested in. And indeed, public capital *was* being attracted: "Since the first allocations in the 2002 state budget, and until today, the government has invested nearly 1 billion NOK in R&D and support functions to develop cod farming in Norway," the 2009 cod plan concludes (RCN/IN/NSRF 2009, 9). When private capital was included, altogether NOK 4 billion had been invested in the cod farming enterprise: in developing cod fry production; researching and experimenting with feed, sexual maturation, and medication; building up cod farms and slaughter facilities; or conducting market research and market work.

At the time of the publication of the 2009 cod plan, it was, however, becoming quite clear that huge amounts of funding were not sufficient. If capital was willing, the cod were not. For as the years had passed, the problems had not been overcome and the situation was far from stabilized. From the 2009 cod plan we learn that the cod had not been as easy to control as the salmon. The model fish and that being modeled upon it did not go as well together as the initial stories of the cod as a "slumbering giant" would have it. The total biomass that cod farming could offer remained too limited and the issues and "bottlenecks" identified by the cod plans remained unsolved. There were huge costs associated with the refusal of the cod to step out of its sexual rhythm. The cod continued to be bothered by diseases, and despite comprehensive vaccination efforts, diseases like vibriosis continued to cause mass deaths. There were also several other, and mortal, diseases, such as *Francisella* infections, which brought on huge losses (RCN/IN/NSRF 2009, 31). And there were problems of resistance to antibiotics due to high applications of such medications (RCN/IN/NSRF 2009, 9). As if this were not enough, the cod continued to swim in their own preferred directions and kept escaping the net pens in the ocean. Such losses were not just economic, however. The escaped cod also became a potential environmental problem, as they could pollute the "wild" cod with their modified genetic material. Even worse, perhaps, was the fact that the escaped cod did not always look like the familiar Atlantic cod. Because of the intensive production methods, many of the domesticated cod grew up to develop "deformities" (RCN/SND 2003, 13). The back and sometimes the jaw were strangely shaped, causing the notion of "monster cod" to enter newspapers and policy debates (Brattland 2013). The cod were literally modified, but in ways that were not planned for as part of the program of innovation.

The series of problems may seem infinite, and the situation rather hopeless. But the version of economization we are dealing with is not one that is easily dissuaded. The innovation paradigm's mode of operation is to identify problems while at the same time rendering them solvable. To do so, it seems, is only a matter of time, research and development, and the required capital investments—with the right qualities. Indeed, in the last cod plan, from 2009, which was also a new ten-year plan, demands for significant public investment were included. Following a long list establishing and categorizing all foreseeable challenges, the 2009 cod plan presents a table designating "the estimated need for public means to resolve the aforementioned bottlenecks through conscientious R&D work in the industry and research environments" (RCN/IN/NSRF 2009, 43). When the different areas requiring assistance were gathered—under the topics Fry/Juvenile Fish, Table Fish, Capture-Based Aquaculture/Cod, Technology and Equipment, Breeding and Genetics, Health and Disease, Environmental Effects, Ethics and Welfare, Safe Seafood, Market and Product Development—the final sum was quite significant. Health and Disease, as well as Market and Product Development, would prove most demanding, with NOK 270 million in "estimated needs" for each. The total amount needed was estimated to be almost NOK 1.5 billion during the period for which the new ten-year plan was intended (RCN/IN/NSRF 2009, 43). Optimism still remained, however, as the plan referred to how then Minister of Fisheries and Coastal Affairs, Helga Pedersen, as late as February 2009, had expressed "sustained faith in farmed cod becoming an important part of coastal value creation." The present ten-year plan would "substantiate these political aspirations" (RCN/IN/NSRF 2009, 10).

Despite all the trouble facing the aspiring cod farming industry, words like "fiasco," "failing," or "futile" are not employed. Instead, it was simply pointed out that in "2008 there are few of the farming companies that make money." During a start-up phase, this was natural, it was stated, but "in the long run we depend on achieving profitability" (RCN/IN/NSRF 2009, 9). This was not, however, the way things went. The combination of too low cod prices and reduced willingness to take risks created, as it was formulated, "challenges for the industry." The future was therefore far from exclusively bright—the expectation now was "reduced growth in the short term" (RCN/IN/NSRF 2009, 9–10). The long term was also paved with challenges, with regard to market prices and with regard to resolving the problem of high production costs. Again, this was connected to diseases and to the difficulty of controlling the cod's premature sexual maturation in captivity.

LIQUIDITY SLAUGHTERING AND SPECTACULAR FAILINGS

In the end it became evident that the entire cod farming industry was collapsing. In 2010, production levels plateaued, before dropping rapidly. In the next three years, the industry experienced a veritable landslide of bankruptcies and company liquidations (Nævdal and Hovland 2014), a process of dismantling companies and selling remaining assets. This also included so-called liquidity slaughtering, whereby all the cod that had been put into net pens, irrespective of their state or size, were slaughtered and sold.[2] The term "liquidity slaughtering," is strikingly descriptive of how the biological and the economic —life and capital—can come together in a single act. The practice of co-modification is also one that works through failure.

By the end of 2014, there was not a single farmed cod left in the sea—that is, except for those that had escaped. As noted by a cod farmer and investor when describing the years leading up to the 2010 collapse: "It was an interesting time, but it was also totally ruin."[3] There were problems with the cod biology, but amid these problems, the market price for cod dropped significantly, largely because quotas for fishing cod were significantly increased. "The market was overrun with cod," as the farmer and investor put it. "The prices were terrible and the companies that were established would need to bring in massive amounts of money to survive. Then, the 2008 financial crisis hit. The capital market 'dried up' entirely and the whole industry went under."[4]

Rather than a new start, the decade from 2000 and onward seemed to repeat many of the earlier experiences from the 1980s, when the cod had, in much the same way, proved resistant to efforts to successfully farm and fit it into the market. Once more, it failed, and spectacularly at that. Yet to narrate this as a sad end is not necessarily in line with the version of economization we have denoted as the innovation paradigm. For instance, and recalling the paradigm of innovation as it was laid out in public policy, it was here emphasized that it was important to also be aware that enterprises that tried out new ideas, and had not succeeded, contributed to increased innovation in the business sector. Entrepreneurs that had failed had often acquired valuable competence, and such competence had to come to use in new projects (St. prp. 2002–2003, 21). However, we would add, when the enterprise in question is that of dealing with living creatures, such as tiny cod fry or grown and spawning cod bodies, the merits in learning and failing—spectacularly—are perhaps not of equal worth or value.

Well outside the document sites of the cod plans, in an interview conducted almost ten years after his own company was liquidated, another former cod farmer and investor tells a story of how fish were put into net pens that normally would not have been put into production.[5] In the early phase, when things were moving slower, he recounts, the cod fry producers were doing a lot more quality grading. But when the businesses were set to scale up and grow large, "everything" was in demand, and everything was put into the ocean, including cod fry of poor quality. "In the end," he states, "it turned out that about 30 percent of the fish that were put out into the ocean should never have been put out. Those 30 percent were bad fish, weak fish, fish that got sick and infected the others." Then came the summer of 2010, an unusually hot summer. Heat makes the fish more prone to disease, and that summer the disease *Francisella* hit the cod hard. The largest, and most valuable cod were the ones that died first, but soon, in some of the net pens, just about all the fish got sick. In the worst period, they would remove between 3,000 and 5,000 dead fish from the pens every day. Then a new shift of workers would come in, working through the night to get rid of the dead fish. "The biology was not in place for running things at that scale," the investor and farmer states. "The idea was to build things up fast, get listed on the stock exchange, earn their money back. It was not about how good the quality of the fry was, only about how much you could get out of it."

INNOVATION AS A VERSION OF ECONOMIZATION

The paradigm of innovation came into prominence in precisely the years when reinvestment in the cod was happening—pushing it forward, yet again, as a promising species. Tied to the agencies of Innovation Norway and the Research Council of Norway, an innovation strategy for the farming of cod was developed, taking the cod on board as an object of innovation; a thing to be researched, coordinated, invested in, and co-modified with systems of production and market demands. In tracing how the cod is brought into innovation, our objective has simultaneously been to analyze the innovation paradigm as a distinct version of economization.

Following the cod plans closely has enabled us to discern the innovation paradigm as a version of economization. In other words, they are sites for "acting out" the very innovation paradigm. By acting as coordination devices they enact the dynamic approach that we observed to be at the core of the innovation paradigm. Key to this is having actors come together, to learn and fail in shared constellations of problem solving—or in other words, co-creation. Contrary to the neoclassical economics, critiqued in the public policy put forward to the Parliament, this is not primarily about describing situations of or detecting market failures. Rather, the plans and the version of economization they are involved in, are about realizations: they are set out to *be* performative, to realize values and act out an innovation economy. Its mode of ordering is to have actors, elements, and, in our case, the cod too, be "in it"—together—the overriding value being coordination and cooperation. The cod plans, approached as sites of innovation as well as tools of valuation, are then not so much formal plans at all. Rather, they are tools that value and underpin coordination and work as tools for making "things" come together. In this respect, they resemble what business models do in the analysis of Doganova and Eyquem-Renault (2009); they not only present a plan, or a model, but work to draw the audience into innovation and attract investors to the project and investment in question. The innovation paradigm is then of a quite peculiar format and there is more to this than, as Dewey (1927) would perhaps put it, being bereft of inquiries and problematizations: As a version of economization the innovation paradigm is less about conflicts, principles, or interests that compete, groups or actors with viewpoints that differ, assumptions that clash, opposing objectives to choose between—winners *or* losers. A key trait of the innovation paradigm is rather to internalize. It takes what is outside *inside, into* innovation as elements to work on, as problems to solve, hindrances to overcome, and "obstacles" that must be surpassed.

The version of economization that is acted out works by a double operation. First, there is capitalization. This implies approaching entities from a particular angle—seeing them *as* capital, turning things and constituting them, as capital (cf. Muniesa et al. 2017). Included in this capitalization, we have argued, is that of attracting capital with the right qualities and to qualify capital—as being good, patient, willing, and risky. Big capital is turned into a precondition to realizing the innovation that is in demand. This is a major shift from the version of capital that dominated the scene in the early years of aquaculture, where non-local capital was something to be kept at a distance. Second, there is marketization. We have shown how this not only implies turning entities into market objects, or things into commodities. Marketization operations are about carefully modifying markets to act in accordance with consumer valuations as they are rendered visible through market research, and then next seeking to co-modify these with the qualities of the thing-turned-commodity. In this way, both markets and the cod are rendered a modifiable material. Our analysis has been examining such operations of capitalization, yet also intense operations of market work, and has discerned how they, together, are entangled in what we identify as co-modifications. What our examination furthermore suggests, is that this double operation goes to the core of innovation as a version of economization. Such capitalization and marketization operations are co-dependent and co-modify one another, this chapter has shown, and it is this aspect of innovation that produces the innovation version of economization as such an all-encompassing endeavor.

Surely, when pursuing the above analysis, our objective has not been to argue that innovation, here at the entrance of the millennium, is something entirely new. It suffices to say that one of the most prominent contributions to innovation theory, by the Austrian economist and politician Joseph Schumpeter (1883–1959), was written already at the beginning of the twentieth century (Schumpeter [1934] 2008); that the study of innovation was quite firmly established as a discipline already in the 1960s, most notably by way of scholars like Christopher Freeman and the Science Policy Research Unity SPRU in the United Kingdom; and that demands for an innovation policy already in the early 1980s were making their mark on research and development (R&D) policies (see Gulbrandsen 2011). Moreover, and it almost goes without saying, that innovation in the very straightforward meaning of producing and doing something new is a key aspect to economic, social, and technological change quite generally. Our take has also not been to analyze an economy of innovation in the form of a theoretical exposé, for instance by outlining the tradition of Joseph Schumpeter (1883–1959) or with delineating the differences (and also similarities) from neoclassical economics and the critique of economists' equilibrium models or other models (Elliot 2008). Taking the innovation paradigm as our object of analysis, we have instead sought to discern it through grasping what it does, in practice and in action.

FROM LIMITS TO GROWTH TO INNOVATION

In *Life as Surplus*, Melinda Cooper (2008) points out that while other new trends and schools in economic thinking have been subjected to extensive and critical examination, the innovation economy has not received the same attention. This is despite the fact, Cooper argues, that it is the innovation economy that represents the true neoliberal economic turn. Starting out from the *Limits to Growth* report of 1972 (Meadows et al. 1972), produced by an MIT community of researchers who predicted that the earth's resources were at risk of depletion and exhaustion, Cooper turns to the critique and criticisms that followed in its wake. She shows how the promises of biotechnology were to work toward internalizing nature into the economy and, consequently, to overcome the idea that there were limits to growth (Cooper 2008, 18). Cooper's analysis thereby connects the issue of biotechnology to the debate on the "limits to growth" approach. Rather than staying with what she frames as a neoliberal turn, her analysis could just as well have addressed the issue of the innovation economy more directly, as innovation scholars were in fact among the most vocal critics to the arguments and hypotheses put forward in *Limits to Growth*. The prediction that there were physical, natural resource limits to growth, innovation scholars argued, underestimated the possibility of technical progress (Cole et. al 1973; Freeman 1973). In this framing, the issue was reframed from inventories over available resources, to that of enabling technological change and innovations (see also Cassen and Cointe 2022). In an innovation context, resources were not scarce in a definite sense. Rather than being something given, they were something to be worked upon, modified, and transformed.

In turning to this chapter's conclusion, what we suggest is that the aquaculture enterprise is particularly apt for grasping the innovation paradigm and capturing how the potential in "the bio"—the cod fish and its natural environment—becomes a challenge, not with regard to limiting growth, but as something to be worked on and overcome. Yet, as the above analysis has shown, this should not be understood exclusively within a frame of technology or technological change, but more overridingly as about how nature is *taken into* innovation. Nature, our analysis suggests, is transformed from being predominantly situated outside the economy, holding limits and worlds of its own, to becoming nature subjected to innovation. In other words, nature is taken into innovation as challenges to be handled and worked on. There are as such no limits to growth, in fact, no limits to modifications, only problems to be overcome. This is not to say that environmental and ethical problems are not addressed and deemed relevant to manage or to care for, but rather that solving them becomes a precondition for further growth (Reinertsen and Asdal 2019).

Starting out in a similar place as Cooper, but pursuing a different path of analysis, we have shown that just as much as the innovation paradigm is about the promise of technological change or progress, and about how technological change is the way to transgress or surpass limits to growth, the innovation paradigm is about transgressing limits through the double endeavor of capitalization and marketization. This is an operation enabled by an alignment of forces where actors, human and nonhuman, are coordinated and *co-modified*. To be sure, such operations can indeed, and quite spectacularly so, fail, the cod farming enterprise being a case in point. In the innovation paradigm, however, this does not represent an end-point, but a learning opportunity in seeking out new ways, new economies, and new ensembles of dynamic and coordinated actors.

6 PRICES FOR COLLECTIVE CONCERNS

It is March and well into the cod fishing season, but so far catches have been low. January and February have been stormy, the boats have been forced to stay ashore, and the large influx of cod moving inshore has been late. Now, however, the *skrei* journeying from the Barents Sea have arrived. Millions of big, fat cod, ready to spawn, easy to catch. So, as the wind gives way to crisp winter weather, calmer seas, and blue skies, the coastal fleet goes out in full force. Small traditional boats and larger, industrial vessels, side by side, set their fishing gear to work. Hand lines, long lines, gill nets, and purse seine nets occupy the sea. Meanwhile, on shore, in the town of Myre, one can sense the anticipation of the work that will follow once the fleet returns. Apart from a baby stroller parked outside the supermarket, the high street is abandoned. At the local café, the only guests are a group of pensioners, who eagerly engage in conversation about how much cod which boat landed last night, followed by speculation about who will do well today. Hemmed in by tall and sharply cut mountains, but facing the open ocean, Myre is home to about two thousand people—a small population, but contrary to most coastal towns, it is a community in growth. Lining its horseshoe-shaped harbor are the tall silos of a fish-feed factory, a state-of-the-art fish processing plant, and altogether three large fish landing stations. The competition between the landing stations to secure the supply of fish is hard, but as put by the owner of one of them, "We cooperate when we can, compete when we have to" (Endresen 2016).

Of all the fishing communities along the coast, Myre is one of the places—often *the* place—where the most cod is being landed (Hansen

2020). That somewhat sharp, salty, tangy, unmistakable smell of fish on the docks, the saying here goes, is the smell of money. And right now, as the cod fishing season is about to peak, that smell is strong. But how, exactly, is fish made into money? How is it transformed from being something that simply exists, swimming about on its own command, to becoming a "commodity," a thing with a market and a monetary price affixed to it?

To investigate these questions, we start our study not in the "classical" market situation where commodities meet their consumers. Instead, we turn to commodity valuations that take place in production processes and in preparation for this "grand finale," one could say, of a commodity's exchange at the market. In doing so, we answer to a challenge posed from within both economics (Nicholas 2012) and valuation studies (Vatin 2013), that when studying markets, one must also consider production. As argued by François Vatin (2013, 40), a link should be established between "a theory of production and work, and a theory of the market and value." In our case, making such a link enables us to pin down how the production of cod commodities intersects with and influences how their value is considered and thereby the process of attaching a price to them (see also Asdal and Cointe 2021). Our sites of exploration are interviews and observations conducted at the physical sites of the fresh cod industry. This includes following the cod throughout a day and night at one of Myre's fish landing stations, but also going out at sea, as a fisher hauls his catch onboard. We enter the offices of fish exporters, the Norwegian Fishermen's Sales Organization, and the Norwegian Seafood Council, and follow along the route of the cod as it is transported to its export markets.

Our analysis of the fresh cod industry draws on a series of relatively recent works within valuation studies and social studies of markets (Çalışkan 2010; Doganova and Rabeharisoa 2022; Reinecke 2010). As we explain in further detail below, these show that far from being an abstract calculation dependent solely on the market forces of supply and demand, pricing is a complex social and material practice. In our use of these studies, however, we make a twist: pricing is a complex social and material practice, we agree. In fact, an important contribution of this chapter is to demonstrate a series of rich, hybrid, and complex organizational forms of pricing that we propose to approach as "valuation arrangements." However, pricing is not exclusively a social affair. As our examination will show, the cod also actively partakes in its valuation and acts on and interferes with the formation of prices. The arrangements and tools involved in valuing cod, including pricing it, depend on various co-modifications whereby working on cod bodies is also about working *with* them. Not a passive entity, this tells us, the cod is one that must be worked with every step of the way to the market. The cod is an entity whose propensities and affordances one must know the most intimate of details of for it to perform well as a commodity. The chapter shows this by examining three quite different valuation arrangements. First, the so-called minimum price that fishers are guaranteed when landing their catch at the docks, a pricing procedure that is the result of a protracted political struggle, steady negotiations, and fine-grained modeling work. Second, the close to iconic and ideal-typical market exchange situation that exists between fish landing stations and fish exporters and that is driven by a volatile balance of supply and demand as well as by a need for speed. And third, the Skrei quality brand, a valuation arrangement that involves strategies for enhancing the market value of the cod through branding, patrolling, and qualifying the cod for premium value. In all these three valuation arrangements, we show, the cod actively shapes and co-modifies the conditions of its valuation.

CALCULATIVE SPACES AND VALUATION ARRANGEMENTS

In engaging with the question of how and by what means the price of a commodity is determined, we raise a question that has occupied economic theory since its very inception. In economics, the neoclassical school of microeconomics has been particularly influential and has conceptualized this as a question of scarcity and competition between market actors (Çalışkan 2010; Nicholas 2012, 458). A key concept here is equilibrium prices, equilibrium being achieved when the demand and supply of a market balance each other, and, as a result, prices become stable. The details of how this is best explained, or calculated and modeled, are widely debated, both within and outside the school of neoclassical thought. And yet the idea that the monetary value of a commodity is always determined in markets and by the relationship between supply and demand remains very strong, in economics and beyond. In contrast to, often also in opposition to formal and abstracted ways of considering market exchange, economic anthropology and sociology have produced a wide variety of accounts that, like ours, rely on qualitative description to show how market exchanges are far more complex than what the models and laws of economics allow for.[1] Prices, this literature has long argued, do not result from a "natural" encounter between market forces of supply and demand, but are part of a wider social practice (Doganova and Rabeharisoa 2022, 3). Significant academic contributions include anthropological debate on gift economies (Malinowski 1922; Mauss 1954; Yan 2013), as well as prolonged debates within sociology on the so-called embeddedness of markets, a concept that speaks to the influence on markets by the wider social, cultural, and political dynamics in which they are situated (Granovetter 1985). In his ethnography of the world's largest fish market, Tsukiji in Tokyo, Theodor C. Bestor (2004, 13–14) argues along related lines of reasoning, writing that: "Markets reflect and generate cultural and social life in wider structures of social life."

In a sense, it is much easier to critique the reductionism inherent to the abstractions of economic theory than it is to argue against the empirical detail of qualitative studies. Yet, as has been argued from the viewpoint of social studies of markets, the challenge of ethnographic accounts is that, rather than tackle head-on the problem of how economic calculations are made, they tend to drown it in details and descriptions. By denying any particularity to economic behavior, Callon and Muniesa (2005, 1230) argue, ethnographic accounts often fail to describe how calculations take place. Markets become "everything and nothing," which in turn means that the nature of practices that are uniquely economic—in our case, putting a price on cod that allows it to be exchanged for money—slips away from analytical attention. To solve the problem of analysis being either too abstract or too rich, Callon and Muniesa propose, one must attend to what they call "calculative spaces" and "forms of calculation": "An invoice, a grid, a factory, a trading screen, a trading room, a clearing-house, a computer memory, a shopping cart—all those spaces can be analyzed as calculative spaces, but all will provide different forms of calculation" (Callon and Muniesa 2005, 1231). And surely, social studies of markets, including the closely related field of valuation studies, have in recent years done much to open up such spaces and forms of calculation to analytical scrutiny (see, e.g., Beckert and Aspers 2011; Çalışkan and Callon 2010, 2009; Doganova and Rabeharisoa 2022; Frankel 2018; Guyer 2009; Pallesen 2016; Reinecke 2010; Reinertsen and Asdal 2019). A key term in many of these works is "market devices," a term widely used in social studies of markets to describe the "material and discursive assemblages that intervene in the construction of markets" (Muniesa, Millo, and Callon 2007, 2). More recently, scholars have also begun to explore the notion of "valuation devices," defined by Liliana Doganova and Vololona Rabeharisoa (2022) as ensembles of narratives and calculations that partake in the construction of markets and in the enactment of values in the economy (see also Doganova 2019; Doganova and Muniesa, 2015; Pallesen 2016). The notion tools of valuation (Asdal 2015b, 2016) that we work from in this book, is put to work, in this chapter, to explore how prices can also be considered, more broadly, as tools that work across the sites of production and markets, but also other spheres, like the political; that also work through other modes of valuation than narrative and calculation, such as the minimum price explored in this chapter; are material *and* semiotic, and whose working is inextricably tied to valuation arrangements in and beyond market work, thereby also serving what we consider in this chapter as more-than-market and collective concerns.

Social studies of markets and valuation studies have opened the issue of price to a broader set of concerns than that captured by economics or economic sociology. Most importantly, social studies of markets and valuation studies have worked to show that pricing is a practice that is realized by a series of material and semiotic means. Pricing is a social practice in the sense of there being a range of actors, rationales, and struggles involved. And it is a material practice, in the sense of there being a range of material devices or tools involved, and which often work in combination with narratives, symbols, and, more generally, semiotic dimensions. Our intervention to the study of prices is closely connected to these material-semiotic tools of valuation and the valuation arrangements of which they are part and that demonstrate the more-than-markets elements to pricing. The main contribution of this chapter, however, is another. Most importantly, we show that the valuation arrangements we examine are set up to handle how also the commodity in question—cod destined for fresh fish markets—actively partakes in and co-modifies its valuation. The material dimension to pricing, we show, is not only about the tools and arrangements by which the commodity is priced and valued but also about the material agency, affordances, and properties of the cod commodities we study.

Our observation of the cod as being an active, *not* passive, commodity runs counter to the claim put forward by Çalışkan and Callon (2010, 5) that the pacification of goods is "an essential property of the regular functioning of markets." Asymmetrical agency—between the *active* entity performing the valuation and the *passive* entity being valued—they further argue, is a precondition for the transformation of a "thing," be it a material object, a service, or a living being, into a tradable good. Is it possible, Çalışkan and Callon (2010, 5) ask, to conceive of a market "in which goods are authorized to destroy this asymmetry of their own initiative and to contribute multiple suggestions of their own value or that of the agencies trading them"? To this, their answer is a "resounding no"—ours would be a firm "yes." For as this chapter moves to show, the cod asserts itself as a commodity that must be handled in quite specific ways. It is an entity that enters the role of commodity not passively, but in ways that co-modify how it can be worked on and imbued with value.

The sections that follow examine three distinct valuation arrangements of pricing. We start out by showing how, in the sites and practices of pricing cod commodities, "more-than-market" elements can come into play—not only as a societal "influence," as implied by embeddedness theory, but as part of the calculative practices by which prices are set. First, and in the case of what is called the minimum price system, we show that political imperatives of sustaining communities and distributing wealth along the coast can be key to how prices are calculated and determined. This valuation practice is the outcome *not* of markets or market dynamics alone, but of protracted political struggles to ensure that the price fishers receive when landing their catch is sufficient to sustain their livelihood. Second, and turning to the supply-and-demand-driven market situation that arises when the cod is resold from the fish landing station, we bring out how cod commodities are far from passive, but actively co-modify the economic practices they are part of. The cod, "in the flesh," asserts itself and directs how its valuation can be conducted, much by directing the practices and speed at which valuation operations must be conducted. Such cod assertions, we show, are evident both in the landing station's production and trade of cod commodities *and* in the third case the chapter examines—the valuation arrangement of the Skrei quality brand and the various tools of valuation that it relies on. In social studies of markets, such work has previously been described, as touched on above, as rendering goods passive, and as essential to them being exchangeable in markets (Çalışkan and Callon 2010). This has furthermore been explored through what the field identifies as the interlinked processes of "qualification" and "singularization" of the good (Callon, Méadel, and Rabeharisoa 2002; Callon 2005; Musselin and Paradeise 2005), the aim of this being to "establish a constellation of characteristics, stabilized at least for a while, while they are attached to the product and transform it temporarily into a tradable good in the market" (Callon, Méadel, and Rabeharisoa 2002, 199). Here, we bring out how the qualification of cod as a commodity also hinges on its active participation in such processes. Valuation operates not only by more-than-market concerns and means, we therefore argue, but also by more-than-market agencies. Furthermore, and building on the notion of "markets for collective concerns," coined by Christian Frankel, José Ossandón, and Trine Pallesen (2019) to describe how markets are increasingly put forward as particularly well equipped to correct market failures (for more on this, see chapter 7), we suggest that one can also consider *prices* for collective concerns. For as we now turn to examine more closely the first of the chapter's three valuation arrangements—the minimum price system—we find that the relations between commodities and their pricing are governed by not only market concerns but also collective concerns that influence the very practice of price calculation.

PRICES AS MORE-THAN-MARKET VALUATION ARRANGEMENTS

Walking and talking, the manager of one of the three fish landing stations in Myre ascends the stairs from the factory at the ground level to the upper floor offices.[2] His conversations on the phone are quick and effective, but as soon as one call ends the phone rings again. "I have to get this," he apologizes and gestures toward his office, "it's a buyer." From the office there is a panoramic view of the Myre harbor and the inlet where boats pass to get to and from the fishing grounds. Putting two mugs of coffee on the table, the manager sets his phone to silent and takes a deep breath. It is now late in the evening, and it has been like this all day. "You start by making a few phone calls in the morning, to check who's buying what," he explains. "Or the buyers call you, usually they call you, and want fish. Then you need to, well, try to guess how much fish you'll get in that day, considering the boats that are out. You don't really have a clue, so you just, well, make a guess."

Buying and selling fish, making deals with fishers on their way toward shore, the hulls of their boats filled to the rim with freshly caught cod, but also making deals with fish exporters waiting to fill up truckloads of fish that they will resell in foreign markets, the manager of the fish landing station is working within two distinct valuation arrangements for the setting of cod prices. One of these, which we return to the details of below, is in fact quite similar to the market situation envisioned by neoclassical economics, in that prices are here largely determined by the balance of supply and demand. The other valuation arrangement, which we will start out to delineate and analyze here—the minimum price system that regulates the firsthand trade of the cod—is very different from this. The notion of "firsthand" trade stems from how the fish, until it is captured, is the property of no one. Once it is hauled onboard a fishing vessel, however, it becomes the fisher's property, and so when it is sold to the landing station, this newly constituted property changes hands for the first time. In many of the fishing communities along the coast, fishers have only one option of where to land their fish, but in Myre there are three large landing stations to choose between. Some days, the competition between the landing stations to secure enough raw material is rife, giving the fishers a strong bargaining position when they land their catch. On other days, when fishing is good and many boats are out, the exact opposite situation can arise and the supply of fish drives prices down. Still, while there are no legal limits to how much the landing station can pay for the firsthand purchase of fish, there is a bottom, and this is where the minimum price comes in. This is a price that holds a long history of struggle between fishers and fish buyers, but also one that, interestingly, is set by valuation arrangements and tools of valuation that work toward consensus; that look not only toward the market the fish is moving to, but also toward the pricing of the past; and which, taken together, work to value the coastal communities of which the cod is part. This is, in other terms, a quite specific version of economization, one that entails more-than-market valuations and a pricing for collective concerns.

The firsthand transactions that take place between the fishers and the landing stations rely on a system of fish sales organizations. These are owned by the fishers and are, with the exception of the sales organization handling pelagic fish, like herring and mackerel, organized so that each of them controls the firsthand sales in one out of four fishery regions.[3] The organization that covers the area from Møre in the northwest to Finnmark in the northeast, which is also the fishery area where most of the cod is captured, is the Norwegian Fishermen's Sales Organization (Råfisklaget).[4]

The system of fisher-owned sales organizations is the result of a protracted political struggle to establish a so-called minimum price for firsthand sales. This struggle, Gunnar Grytås (2013) writes, was spurred by the outbreak of World War I in 1914, which put an end to a long good period for the Norwegian fisheries. Times of prosperity had incited optimism among the fishers, many of whom had taken out bank loans and invested in larger and better boats. The war, however, disrupted the fish trade greatly. It closed important markets in Southern Europe to Norwegian exporters and caused strong inflation and currency fluctuations. When the war ended four years later, it left the European markets with significantly weakened purchasing power and new import barriers. In addition to this, many of the countries that had previously bought fish from Norway now opted to become self-sufficient, sending newly purchased trawlers up along the Norwegian coast and all the way up to the Barents Sea. The Norwegian fisheries consequently entered a long period of sharply falling prices and bankruptcies. From 1914 until 1926, the price of cod fell, consistently and fast, from NOK 0.138 (13.8 øre) to NOK 0.062 (6.2 øre) per kilogram. By 1932, the price had only increased incrementally, to NOK 0.073 per kilogram of cod exported from Norway.

The decline of fish exports led to rapid poverty growth in the many fishery-dependent communities along the coast (Grytås 2013). Responding to the destitution caused by this and seeking to improve their position in relation to the fish buyers and exporters, fishers began to organize all along the Norwegian coast, many of them in close alliance with the growing labor movement. Several local and regional fisheries organizations were established, and in 1926 the Norwegian Fishermen's Organization (Norges Fiskarlag) emerged as the one national organization that was to represent the interests of more than 100,000 fishers. Not long after, the suggestion of a minimum price system was put forward. The idea was to establish a scheme that could guarantee the fishers a set price per kilogram of fish landed and, equally, guarantee that in cases where the market price achieved by the exporters was below the minimum price, they would be compensated for this by the state. The minimum price system consequently emerged as a distinct valuation arrangement, valuing, on the one hand, the livelihood of the fishers and their communities and, on the other, the risk taken by the fish buyers when paying a minimum price. Handling both market and "more-than-market" valuations, the proposed scheme was to be a tool not only for setting the price of fish but also for resolving political conflict and for a more viable distribution of the wealth generated by the fisheries.

The proposition for a minimum price system was taken up by the Norwegian Parliament. Here, the work dragged on, however, leading to growing resentment among the fishers. Pressure was being asserted from many directions, and by way of quite different strategies the fishers managed to transform policy (Hersoug, Christensen, and Finstad, n.d.)—the most spectacular initiative being made by the fishers of Vardø, who in September 1937 took matters into their own hands (Nielsen 2012, 68). Situated in the northeast of Norway, Vardø was at the time one of the most important landing sites for Norwegian fisheries, with cod being the most significant species. Through the actions of the fishers, the harbor that normally shielded boats from the winds and waves of the Barents Sea was transformed into a hot political battleground. Pulling thick chains across the mouth of the harbor, the local fishers' organization put an effective stop to the fisheries, a stop that would later be described as a "battering ram" in the work to implement a minimum price system. It was also the first strike to take place in the history of Norwegian fisheries, and similar actions soon took place in many other coastal communities. Responding to the pressure caused by this, the government agreed to appoint a "fast-working committee" to assess the principles of a minimum price agreement, and by June 1938 the Raw Fish Act (*Råfiskloven av 1938*) was passed in Parliament.

Following the new Raw Fish Act, it became illegal to conduct firsthand sales of ocean-captured marine fish outside the fish sales organizations (Grytås 2013). The right to set the minimum price was firmly put into the hands of the fishers, the intention being to secure the fisheries as the economic backbone of coastal communities. Since then, the act has undergone several major revisions, the last in 2013 when it was renamed "Act on Firsthand Sale of Wild Marine Resources."[5] The intent of the initial act, however, still stands strong. As stated in paragraph 1 of the current version, the purpose of the act is to contribute to a sustainable and socioeconomically profitable management of marine resources living in the wild.[6] This purpose is further enforced by the Ocean Resources Act (*Havressursloven*), which underlines that such management should contribute to securing employment and settlement along the coast.[7] "The marine resources of the wild belong to the community of Norway," the act firmly establishes—an affirmation that, recalling from chapter 2 the negotiations that took place in the 1960s and 1970s over the exclusive economic zone, ties in with national interest as well as with the historical rights of users of the ocean commons. To have a population dependent on and actively using the ocean is closely tied to the nation's legitimate claim to the control over and exploitation of ocean resources. To secure the viability of coastal communities is to secure strategic national interests.

Grounded in law and by a long-standing political principle of user-based rights to the ocean, the minimum price system is one where the price of fish is intimately tied to the more-than-market and, indeed, collective concerns of securing coastal communities through their benefit from the fisheries. It is a valuation arrangement and tool of valuation that values more than a market good. This entails that the notion that prices are primarily "market devices" (Muniesa, Millo, and Callon 2007, 2) is too narrow for the purposes of our analysis. It simply does not capture the more-than-market aspects to the minimum price system that, while decisive to the determination of the cod's firsthand price, are not reducible to a market dynamic. Prices are quite evidently—to the point of being an icon of the market economy—market devices, but they are also more broadly tools of valuation that can work toward handling collective concerns, encapsulating—in our case, both political concerns and market prices. The above also speaks to what we have suggested thinking of as "versions of economization," a notion that underlines how processes of rendering something economic may happen in very different ways, by different means, and toward different ends. Here, with the minimum price, what we encounter is a version of economization wherein the setting of a price is as much a politization as it is an economization. It is a more-than-market form of pricing.

THE MANUFACTURING OF A MINIMUM PRICE

Before the implementation of the minimum price system the fishers had been at the mercy of the landing station owners, but now the tables had turned entirely: the power to set the price for firsthand sales was placed with the fishers. However—and importantly—since the very institution of this valuation arrangement, this authority has been exercised by organizing negotiations between the fishers and the fish buyers, the first such meeting taking place the day before Parliament passed the Raw Fish Act of 1938 (Grytås 2013). So instead of the fishers simply naming a price, the Norwegian Fishermen's Sales Organization has worked to gather representatives not only from the fishers' organizations but also from organizations that represent the landing stations and fish exporters, inviting them to negotiate a minimum price that can be acceptable to all parties. If agreement cannot be achieved, the sales organization determines a price for the next week or so, and continues negotiations until a consensus has been reached. In addition to being a tool for the valuation of the cod as a source of living for fishers and their communities, the minimum price can be considered as a form of *consensus price* that takes on the concern that the firsthand price must be profitable to all parties.

The minimum price system is unique in terms of how it is organized through fisher-owned sales organizations authorized by Parliament to determine a minimum price. Still, there are other types of minimum price arrangements that it can be compared to, such as those of the Fairtrade Labelling Organizations, or FLO (Reinecke 2010). FLO is the custodian of the Fairtrade certification mark, which is used to signal to consumers that the commodities they are buying have been purchased at a "fair" minimum price. Fairtrade products are typically agricultural products (like coffee, tea, cotton, and fruit) that are produced by farmers in the Global South and sold to consumers in the Global North. According to Reinecke (2010, 563), the work of the FLO "makes visible the political confrontation at the point of price determination, notably by providing a social arena for where conflicts of interest between opposing parties are played out." A similar type of description could be given of how minimum prices for the firsthand sale of fish emerged in the 1920s and 1930s, as both these and Fairtrade prices are settled not only in markets, or with the aim of establishing a market price, but are tools of valuation that aim to create a more even distribution of wealth. And yet, comparing the Fairtrade certification mark to the minimum price system of today, they come across as being quite different. For as the primary rationale of Fairtrade is still to create more "fair" conditions for "farmers and workers in poor countries" (Forbrukerrådet, n.d.), the minimum price system for the firsthand trade of fish now works in an entirely different context. The authority to set a minimum price is here established by law and is not reliant on the willingness of engross buyers or consumers to enter "fair" trade relations, nor are Norwegian fishers of today the victims of severe economic exploitation. Depending on their access to fishing quotas, they can earn far above the average wage. Correspondingly, also the minimum price arrangement has evolved. It still works to secure sufficient firsthand prices, but the minimum price has also become a tool that values by following market trajectories and, as we now turn to, by anchoring prices in the past.

The negotiations over minimum prices have at times been very conflictual, with strong disagreement between parties over not only what the price should be, but how it should be reached (Bendiksen 2018, 5). The negotiating parties are often in disagreement on how to assess future developments in markets, and there is often great insecurity about many of the factors that affect the export price. As we examine in further detail below, the international market for fresh cod can change quite fast, but up until 2016 several months could pass between each minimum price negotiation. The market situation could therefore change quite a lot while the minimum price would remain set. The Norwegian Fishermen's Sales Organization therefore took the initiative to supplant the price negotiations with a so-called dynamic minimum price model. With the support of the fish buyer representatives, the model came into force in October 2016, and since then the minimum price has been calculated by use of an equation that factors in the price of fresh cod, frozen cod, and export value:

(80% of the average firsthand price of fresh cod + 70% of the average firsthand price of frozen cod + 60% of average price export)/3 = minimum price.

The dynamic minimum price is calculated every fourteen days and is adjusted only if the new calculation shows that prices have either risen or fallen by NOK 0.25. The advantage of this mode of calculation, Bjørn Inge Bendiksen (2018, 5) argues, is that it allows the minimum price to develop with the market, rather than being dependent on the negotiating parties' ability to foresee "in part very complicated relationships and often unpredictable changes that cannot be influenced or foreseen." The dynamic minimum price model is seen to be less static and better attuned to market developments, which is in a way interesting, in that it is based on prices achieved in the past. Contrary to the much more common method of discounting, and which looks to the future to assess the value of things today (Doganova 2018), the model looks to past prices and price trajectories to calculate the price of today and tomorrow. In more than one way, then, the minimum price for cod can be said to be a price with a history. It is a price set not in the moment, and not to secure the highest possible price, but is a price that is considered adequate. Further, and considering how the dynamic minimum price model firmly places the minimum price on a market-driven trajectory, it can be considered as being geared toward the preservation of predictability and stability in cod markets. It does not represent an idealistic counter-movement to "free" competition but is rather a valuation arrangement that determines firsthand prices by other means than looking to the current supply and demand situation. Still, it is also a valuation arrangement that is dependent upon past prices being considered as adequate by the fishers. If the market situation were to change dramatically, both this valuation arrangement and the tools involved in setting the minimum price are likely to, again, be up for debate and negotiation.

FAST PRICING OF QUICKLY DECAYING COMMODITIES

As indicated by the notion of "firsthand," the minimum price is also just that, a "first price." It is a price that is set to be followed by other, preferably higher prices as cod commodities continue their journey toward the consumer. Contrary to the minimum price, however, the ensuing prices of fresh cod commodities are, as we now turn to, market prices. Not unlike the market situation envisioned by neoclassical economics, they are settled in a situation of market competition and by the balancing of supply and demand. This is a pricing procedure, however, that is also closely related to the affordances of the cod. To fully capture the cod's subsequent pricing we must therefore also consider how the cod, rather than being rendered passive asserts itself as a commodity that must be handled in quite specific ways.

The market for fresh cod consists of several different types of buyers, in Norway and abroad, including the producers of various kinds of dried and salted cod, fish-processing factories, supermarkets, and the so-called HoReCa segment—hotels, restaurants, and catering businesses. Fresh cod can be shipped to markets as distant as those of North America and Asia, but the main bulk is sold to buyers within the European Union, where much of it goes into fish-processing factories in the United Kingdom and Poland. These different categories of buyers can have different demands, both in terms of the product quality and volumes they ask for and in terms of the prices they are willing to pay. The category of buyers that a fish landing station predominantly sells to is therefore quite determinative of its activities, which in turn means that there is quite a lot of variation between the roughly 245 fish landing stations situated in the jurisdiction of the Norwegian Fishermen's Sales Organization. The landing stations vary in size and capacity, in what they do to the fish, and how they resell it. For instance, some of the landing stations are also fish-processing factories, meaning that they process the cod into finished products, packed and ready for the consumer market. Others simply prepare the cod for shipment by gutting, heading, and washing it before packing it into crates. How the cod is resold also varies, as some sell directly to, for instance, supermarket chains and restaurants, while others sell to an exporter who then acts as a link to yet another buyer. Before the fresh cod reaches its consumer, it may therefore have been bought and resold several times. This is also the case of cod coming into the Myre landing station. Here, the fish is mostly gutted, headed, packed on ice, and resold to exporters.

The firsthand minimum price provides a bottom line for the landing station, in that to make a profit it needs to resell the fish at a higher price. When reselling, however, there are no legal limitations to how high or low the price can be. Instead, what is put into play is a rather volatile balance between demand and supply. Cod markets are very vulnerable to changes that influence the purchasing power of consumers—such as those brought forward by an international financial crisis, a global pandemic, or sudden shifts in the world's geopolitical balance—but fluctuations can also have a more temporary character. For instance, if a large Spanish or French supermarket chain decides to have a campaign on cod, they will, as put by one fish trader, "vacuum" the market for cod.[8] Or if bad weather on the European continent is discouraging people from going out to eat at restaurants, demand can drop quite heavily. It is not unusual, therefore, for sellers of fresh cod to keep a close eye on the weather forecast of major importing countries—but also of the fishing grounds. For as explained by the manager of the Myre landing station, the buying and selling of fresh fish is a lot about sensing markets, but it is also about sensing the fisheries. Which boats are out? What are they likely to catch? How much? What is the weather going to be like in the next few days? Can they expect a week of high supply, or will storms force the boats to stay in port?

To demonstrate how he oversees the fisheries, the landing station manager flips around the computer screen on his desk and points to the so-called automatic identification system, or AIS, application. This is a system put in place to map, in real time, ship activity in ocean areas.[9] The manager has zoomed in on the fishing grounds just outside the harbor. The banks are thick with red arrows, each arrow representing a fishing vessel. Zooming in even more, he clicks on one of the arrows, revealing the name of the boat, its status, speed, direction, and depth. Following the same webpage, one can also find a real-time overview of the types of fishing gear in the sea. Red arrows represent gear from line fishing, small blue arrowheads represent drift nets, and purple arrows represent purse seine net fishing. Having assessed the fleet, its size, and activity, perhaps talked to a captain or two to hear whether catches are good, the manager can form an idea of how much fish will be landed on a particular day. The manager is then in a better position to come to an agreement with the buyers not only on the price, but also on how much fish can be offered to them. Sometimes the price is set for the entire week, other times it can be from hour to hour. "For instance," the manager says, "a buyer can call and ask if you have fish that can be ready in two hours. If you do, you sell it. There are all spectrums. Like yesterday, I sold enough fish to fill two trucks, and now we have enough fish, so we'll send that off tonight. That's also to do with efficiency—there is no use in sending half-empty trucks to Europe."

What drives the fluctuations of cod markets varies; the point is that when reselling the cod, supply and demand are decisive factors in price determination. The dynamics of the fresh fish trade are in this respect quite like those described in neoclassical theory. And still, the information available to buyers and sellers to assess the market situation and thereby the market value of the cod is only indicative of how the fisheries (supply) and consumption (demand) will develop in the coming days. It is also a highly volatile market, as it is one that can shift at the whim of the weather or a supermarket chain's campaign. If equilibrium is the correct way of describing the situation in which a price is agreed upon, it is not one that lasts for long. Also in this way, then, the pricing of the cod being resold from the landing station is quite different from the minimum price system. For whereas the firsthand price is stably set in the past and by way of a fixed price model, the prices achieved by the landing station are stable only in short time intervals. Within only days, sometimes even hours, the one and same cod is priced twice, or more, and by way of radically different valuation arrangements and tools of valuation—the minimum price system and in a situation of market competition. Sometimes, the cod's market value is also worked upon and enhanced through other valuation arrangements, like that of the quality brand Skrei. A precondition to valuing and enhancing quality, however, is knowing the cod and how it behaves, while alive, but no less, also after it is dead.

NON-PASSIVE COMMODITIES AND THEIR CO-MODIFICATION

No one is in such a hurry, the saying goes, as a fresh fish destined for the market. On a general level, this can of course be explained by how lowering production time also lowers costs, but in the case of the cod fisheries this is not the full explanation. For if the fish landed at the docks sits too long without being processed, its quality deteriorates so fast that soon it becomes inedible—it will have no market value at all. Correct packing and storage can prolong its so-called shelf life—the number of days when it remains fresh and edible—but compared to canned, dried, or frozen food the shelf life of a cod is as short as that of the mayfly. From when it is caught to when it is consumed, it stays good for about ten days. To the landing station, this means that it is essential to both sell the fish and move it through the production line fast. Preferably, the cod begins to make its way toward markets only hours after it is packed. Price is not only about supply and demand, this tells us, but also a matter of the temporalities set by the biological propensities of the cod itself, its moment of death marking the beginning of a new life for the enzymes and microbes working to break down its meat. Even postmortem, the cod has the capacity to act upon, make demands toward, and co-modify the valuations being spun around it. At the landing station the short life of cod commodities comes to view in how the entire production plant is set up to move as fast as possible. As the landing station manager closely surveys how much fish is bought and (re)sold, he therefore also keeps a close eye on the flow of fish moving through the production lines of the ground-floor factory.

On the day of our visit, one of the machines broke down in the morning, leading to a full stop in the entire production. By early afternoon it has been fixed, but it is difficult to catch up, as the capacity of the landing station is already under pressure. Governmental regulation since 2009 has opened up the coastal fisheries to larger vessels—"large coast," as they are called—and larger boats mean larger volumes being landed at the same time. New machinery has been put in place, but from the long plastic tube that sucks the fish from the hull of the boats and into the factory, too much fish is coming in too fast. The workers, most of them seasonal laborers, are working long hours. Stationed along the landing station's production line, they are cutting and gutting the cod coming in, separating the innards and heads from the cod bodies. Crate after crate fills up, some with heads, others with liver or roe. Speeding around the factory floor, forklifts scoop up the crates and deliver them for transportation. Heads go to a company that specializes in dried cod heads for the African market; the cod roe and liver are sold separately; other innards are ensiled and sent to a feed production company. The remaining cod bodies continue their way down the production line, where they are washed and sorted according to size, before being packed in large wooden crates lined with plastic sheets. Piled up and covered with ice, they are commodities ready to go.

Speed is of the essence to this work, but the landing station is not fully at the mercy of the cod, its deteriorating flesh, and how this corresponds to its market value. Or put differently, the cod also lends itself to other ways of turning it into a commodity, as its lean, white meat is particularly apt for preservation by drying and salting it. "Hanging the cod"—that is, splitting it and hanging it outdoors, upside down by a rope attached to its back fin—is therefore an alternative to selling it fresh. In fact, much of the fresh cod goes directly into the production of such "conventional" products, as they are called; some of the methods for drying cod date back as far as memory goes. Next to the empty boxes waiting at the end of the production line, ready to be filled with fresh cod bodies, are therefore also huge bags of salt, equally ready and able to afford the cod a longer life as commodity.

In this case, our interest in salting and drying cod does not lie in this product segment in and of itself. Rather, it is the opportunity it provides to co-modify cod and markets in other ways that we want to highlight. Here, co-modification is captured not by how cod commodities demand being processed and consumed within a certain time frame, but rather by how they lend themselves to other tools of valuation, tools that work more flexibly and at other speeds. For instance, if, for some reason, it becomes impossible to resell all the fish at an acceptable price, many landing stations have the option of shifting their production from fresh cod exports to hanging. In the case that the need for making such a shift has been foreseen—for instance, by a seller keeping a close eye on the weather forecast in major markets—the production line will have been prepared for this and the landing station can still produce a high-quality product that will achieve a good price. The drawback is that returns come later in the year, as dried and salted cod requires more time to be market-ready. If the inability to resell the cod takes the landing station by surprise, it is also likely to salt or hang the cod, but as operations become a matter of a quick turnaround—for instance, by unpacking and reprocessing entire truckloads of cod that have been prepared for transportation—the quality and, with that, the price that the landing station achieves for these commodities is likely to be lowered.

The production of conventional products provides the landing station with a safety valve of sorts. It allows it to control, or at least influence, the flow of cod reaching the fresh fish market. A production method, hanging is also, in some situations, a tool of valuation that works to enhance market value. There is here a very intimate relationship, this tells us, between what goes on in production and what goes on in markets, a relationship that brings to light how the production of value—or what Vatin designates as "valorization"—is "ingrained in acts of work" (Vatin 2013, 14). Still, this does not mean that work or production is always set up to achieve the highest possible price for the commodities produced. Sometimes, the production–market relationship works in such a way that for the individual actor, the rational thing to do is to produce a lower-quality good, or even sell what is in fact a premium quality cod to lower-paying markets that ask only for lower-quality cod. This is, then, also where the imperatives of what we identified in the previous chapter as the innovation paradigm enter the work of valuing fresh cod commodities. For while the non-maximization of the market price obtained can make sense on the level of the individual actor, it is considered as problematic on a more general, industry level.

A TOOL OF VALUATION, A TOOL OF INNOVATION

Early in the morning a fisher sets his nets. In the dark of night, he returns and begins to haul them out of the sea. A machine does the heavy lifting, landing the cod upon a tray where the fisher untangles it from the thin plastic meshes of the net. Relatively speaking, the ocean is calm, but in wintertime, this far out at sea, the waves are always high. The boat rocks from side to side, the cod slide back and forth in the tray, the fisher untangles cod after cod, cuts their throats and throws them into a tub of running seawater. The Arctic night gives way to dawn, the horizon is colored green, then pink, then the pale blue of day, and the catch is good, the best of the season. In fact, it is almost too good, as the cod being pulled out of the sea are almost too many to handle. Sometimes, a fish gets jammed on its way onto the tray, but the machine continues to pull at the net. More and more cod get stuck, some get squashed, others cut by the pressure from the thin threads of the fishing net. Mostly, however, the cod arrive in neat rows, alive and fervently flapping, in mint condition. Their throats cut while they are still alive, they bleed out fast, leaving their meat unblemished and bright white.

"I could do a top-quality job," the fisher states. "I could pull the nets every two to three hours, to make sure that all the fish I take out are alive and kicking, give each cod top-quality treatment. Cut its throat immediately, let it bleed out for at least twenty minutes, in running seawater, gut it, and lay it on ice slush."[10]

The fisher should know what he is talking about, as he was awarded "Quality Fisherman of the Year" by the Norwegian Fishermen's Sales Organization only a few years back. But even if doing a top-quality job could considerably raise the market price of the cod that he is currently hauling in—as well as extend its shelf life by as much as four to five days—the fisher himself will not be rewarded for doing a "top-quality job." He lands his fish not in Myre, but farther north, on an island outside the city of Tromsø where there is only one landing station. When he lands his fish, he explains, the price is the same, irrespective of quality. A few years back he had an agreement with the landing station that he would receive a higher price for the cod he landed, based on his track record of delivering a high-quality catch. But when the landing station changed owners, the agreement was stopped. It would simply be too much work, it was argued, to determine that the cod delivered at a premium price was in fact premium cod. "For such things to work," the fisher says, "there needs to be trust. The people buying the fish need to trust that I do the quality work properly. And I need to be able to trust that if I do a good job, then I get something in return for it. For me, there are extra costs and a lot of extra work involved in doing high-quality fishing. Without extra pay, there is no point in doing it."[11]

On the part of the Myre landing station manager, however, aiming for premium quality production is not so much a question of trust as it is one of speed and capacity.[12] When the production line is working at full capacity, there is simply no time to put in the extra effort. Also, he underlines, what good quality is is not a straightforward question. "In my mind," he states, "to deliver quality is to deliver what the customer asks for. Subjectively, we can say that 'Oh! To me, that is really poor quality' and still, for that one customer, with his needs, it could be just what he wants. Of course, if the quality does not satisfy the customer, it is not good enough. If you are delivering fish to a top restaurant in France, it needs to be super-top. If there is one little discrepancy, they will complain, even though the quality is overall fantastic. Really, you could send someone the Rolls Royce version of the cod, and still, it is not what suits their needs."[13]

In the market situation where the cod meets the consumer—for instance, a supermarket shelf or a restaurant dinner plate—there is a strong connection between the quality of the commodity and its price. Still, as exemplified by the statements of the fisher and the landing station manager, this price–quality relationship can be quite weak. Indeed, as suggested by a white paper on the competitiveness of the Norwegian seafood industry (St. meld. 2015–2016), but also in several of the other interviews we have conducted with actors in the fishing industry, the work conducted by the fishers and at the landing stations too often results in the cod being of too low a quality for it to be considered as a premium quality fish. That the quality of the cod being captured is not maximized is often problematized as a form of loss, in that producing commodities that yield a lower price makes little sense when working with a finite resource. There are only so many cod in the ocean, so why not maximize the value of every cod captured? Several quality-enhancement measures have consequently been put in place, including the legally binding "Regulations on the Quality of Fish and Fish Products."[14] The most direct measure targeting the quality of cod commodities is the trademarked quality brand "Skrei"—a valuation arrangement that works by signaling to consumers that what they are buying is in fact a premium quality fish, and by enticing fishers and fish landing stations to improve their work and produce a higher-quality cod. It is, as we now move to show, a valuation arrangement that consists of several tools of valuation, tools that, moreover, are also tools of innovation. Also, and pointing back to the argument made in the previous chapter—that co-modification can be considered as the modus operandi of the innovation economy—the Skrei brand is set up to make the propensities and affordances of the cod a key concern for fishers, landing stations, and exporters.

Only if these are considered, with care and precision, can a cod commodity be branded as Skrei and be expected to achieve the extra market value that this yields—an estimated NOK 5–10 extra per kilogram.[15] Here, then, the ordering capacities of prices are mobilized to act on the actions of the entire fresh cod industry.

SKREI: A BRAND, A GUIDE, A STANDARD, AND A PATROL

The Skrei quality brand is exclusive for northeast Arctic cod, the Barents Sea cod stock that in Norwegian are also called *skrei*. The brand is owned and managed by the Norwegian Seafood Council, which registered it as a trademark in 2005, but also involves the Norwegian Fishermen's Sales Organization, with whom the Norwegian Seafood Council collaborates to control that the brand is not misused. The Norwegian Seafood Council is a public corporation owned by the Ministry of Trade, Industry, and Fisheries, and its main mandate is to work on behalf of the seafood industry to raise the values and volumes of Norwegian seafood exports. In chapter 7, we examine its work and organizational structure more closely and will for now focus on its efforts to develop the Skrei brand.

FIGURE 6.1
The logo of the Skrei brand, courtesy of the Norwegian Seafood Council.

The Skrei brand was originally intended for business-to-business trade, but in recent years it has also been used in marketing aimed directly at consumers.[16] The right to use the brand is licensed by the Norwegian Seafood Council to landing stations and exporters, the Seafood Council also provides license holders with brand material like stickers and tags, and, importantly, the *Skrei Guide*. Based on the publicly registered quality standard "SN/TS 9406: 2021–Skrei," this thirty-four-page guideline gives detailed instructions on how the fish must be handled for the desired quality to be achieved. Some of the instructions are directed at the fishers and their handling of the fish upon capture, others at the landing stations where the fish is sorted and packed, or at the exporters, detailing how the fish should be handled during transport. No matter who the receiver is, however, the message from the *Skrei Guide* is very clear: utmost care must be taken in every step of the process if premium quality is to be achieved. The guide works to draw together actors with different roles and interests in the fresh cod industry, thereby doing a type of innovation work that we identified in chapter 5 as coordination.

When captured, the *Skrei Guide* starts out to prescribe, the cod must be cut and bled out while still alive, to avoid discoloration of the meat, skin, and fins by blood. The cut should be made by one single stroke across the cod's throat and aorta, ensuring proper bleeding and avoiding "ugly" neck cuts. The fisher must then let the fish bleed out in running seawater, to remove blood and fecal matter from its flesh. The fish must then be cooled down toward a temperature of 0–2°C and stored in precooled and clean seawater. If not properly cooled, the *Skrei Guide* underlines, the fish will deteriorate faster and its shelf life is reduced. Already at the stage of its capture by the fisher, the propensities and affordances of the cod are taken into and co-modify the value enhancement work of the Skrei brand. It does not stop at this, however, for as stressed throughout the *Skrei Guide*, it is at all stages of key importance to handle the fish gently, thereby avoiding visible damage from it being hit, cut, or squashed. Fish with damage from handling—including damage from the tools used to capture it—do not qualify to be branded as Skrei. Once the cod is delivered to the landing station, further handling includes gutting and heading it, which according to the *Skrei Guide* must be conducted in a manner that ensures "nice" and "even" cuts along the cod's belly and neck. The cod should be packed no later than twelve hours after being captured and must then hold a temperature toward 0°C. The box containing the fish should have a five-centimeter layer of ice in the bottom and the cod should be placed in the box with its belly down; if laid on its side, the ice will cause the texture of the cod's skin to change, making it crumpled. There must also be plenty of ice around the neck cut, as the exposed meat makes it vulnerable to deterioration. Ice touching the skin of its back must further be avoided, as this will cause spotting in the color of the skin. During transport, the fish must be accompanied by the proper amount of ice, to ensure that melting does not occur, thereby avoiding blood-stained melting-water seeping out before delivery. Finally, a tag with the image of the Skrei brand must be placed correctly on the fish body, the placement of the tag depending on the category of Skrei product.

In the *Skrei Guide*, the written instructions are accompanied by photographs of cod in various stages of the process—green smiley faces signifying examples of correct handling of the cod, sour red faces signifying incorrect examples. Also, and underscoring the importance of the instructions, the guide contains several prescriptive slogans, such as: "This is the flagship of fresh fish for consumption"; "One should not *think* that the quality is adequate, one shall *guarantee* that it is"; "Good bleeding, fast cooling, and good cleanliness together provide the foundation for optimal perishability"; and "Nice fish—well iced!" Not an entity with passive agency, this tells us, the cod must be worked with every step of the way to meet the requirements of the Skrei brand. Put differently, this is an entity whose non-passivity one must know, understand, and accommodate, in order for it to perform well as a commodity. One must let the cod modify one's practices and do so to the point of perfection. Throughout the work involved in this valuation arrangement, including that of policing the brand, acting with and accommodating the cod's affordances is a key component. Enhanced value, we learn, can only be accomplished in a co-modification process where market value is achieved in intimate interplay with the cod's own propensities and affordances.

While posing strict demands, the *Skrei Guide* is not set in stone. It can be revised by way of established procedure, which it was when the Norwegian Seafood Council got feedback from the industry that requirements for fillet products needed to be more accurately specified.[17] Upon receiving this feedback, the Norwegian Seafood Council invited all who have an interest in the Skrei brand to participate in the work to revise the standard on which the *Skrei Guide* builds. It set up a working group that discussed different issues and tried to reach a consensus on what the revisions should be. The working group's suggestions were then sent out to the entire industry, so that everyone could comment. These comments were subsequently handled by the working group, which again sought to make a proposal that all could agree on. Not a consensus price, but a consensus standard is thereby established, a tool of innovation that requires both industry agreement and coordination with market feedback to be effectual. The market and its demands are with this taken directly into the co-modification work of the Skrei brand.

To ensure that the industry follows the Skrei standard, the Norwegian Seafood Council uses a variety of tools of valuation, including the abovementioned *Skrei Guide*, but also, and perhaps more spectacularly, there is the active policing of the Skrei brand by the so-called Skrei Patrol. The members of the patrol are recruited among employees in the Norwegian Fishermen's Sales Organization, who conduct random checks at landing stations, national and international transport terminals, and supermarkets abroad. If the Skrei Patrol finds that more than 25 percent of the fish from one landing station has discrepancies, the landing station is automatically barred from using the Skrei brand for four days—the time it is assumed it will need to make improvements. As explained by the Norwegian Seafood Council advisor in charge of the Skrei brand, it is their job to "protect" the brand. "For the brand to have value," she states, "the high quality needs to be maintained, from the beginning to the end of the season."[18] There must be a relationship of trust between the brand and the consumer, it is assumed, and if this trust is broken, the willingness to pay a premium price for cod branded as Skrei will soon disappear.

The work conducted to maintain and police the Skrei brand speaks to what social studies of markets describe as the qualification and singularization of goods. Still, it shows, yet again, that for the cod to have a consistently high value it cannot be rendered passive. Rather, one must take seriously its agency and how this can be co-modified vis-à-vis the production and markets of the fresh cod industry. The status of cod commodities, we therefore find, cannot fully be described in one-way terms like "singularization," "qualification," or "standardization." Instead, the arrangements and tools involved in valuing cod, including pricing it, depend on various co-modifications whereby working on cod bodies is also about working *with* them (on this, see also Asdal and Cointe 2021).

POLICING THE MOST PERFECT OF COD

The travel routes of the cod leaving the Myre fish landing station can differ from day to day. Some days entire truckloads begin to make their way to a single buyer—say, a fish-processing factory in Poland. Other days, the trucks leaving the landing station will carry cod for multiple buyers. In this case, their first stop is often the DB Schenker transport terminal in Oslo, which features its own cold storage hall for chilled, not frozen, fish. Here, the cod are unloaded and reorganized into new shipments, before, finally, making their way to their respective buyers. The terminal is therefore considered a good place by the Skrei Patrol controllers to conduct quality checks, making it one of their regular stops and—at the end of the cod fishing season—also the patrol's last stop. The controllers have then already been to a shipping terminal in Padborg, Denmark—where just about all seafood leaving Norway for European Union markets is customs-cleared—and report that they have seen "a lot of fine-looking fish." At the DB Schenker terminal, they are greeted by a manager who knows the controllers well, and immediately lets them into the cold storage hall.

The hall is a neatly organized, square room filled with nearly identical white polystyrene boxes, stacked in rows. Dressed in yellow reflective vests with "The Skrei Patrol" printed on their backs and identical caps from the Norwegian Fishermen's Sales Organization, the controllers immediately start searching for boxes with the Skrei brand on the side. The boxes they find are carefully removed from the stack and placed upon a table. A sticker on the side of the box identifies the exporting company and packing date of the fish, and they routinely begin to photograph the sticker and write down its details on a form. The form also indicates what they are going to measure: the amount of ice and temperature in the polystyrene box, which gives an indication of the fish's perishability; the physical appearance of the cod; whether it is properly bled out; its consistency; its smell; and how it has been tagged. "It's not a scientific assessment," one of the controllers explains, "but this screening gives us a pretty good impression of what the situation in the industry is. We check if the fish have been damaged. Cod branded as Skrei is supposed to be unblemished. There shall be no marks, the fins need to look nice, everything needs to be . . . well, it should be a perfect cod."[19]

The lid of the first box is carefully removed, revealing three cod, headed and gutted, but otherwise whole. One of the controllers examines the neck cut: "Danish Seine fish again," she sighs, before explaining that the Danish Seine fishing method can be very stressful for the fish. It involves fishing from a rather large vessel, using one large net that is cast out wide, then drawn together, pulling the fish into one big bag. "When the fish gets stressed, it pushes the blood into its capillaries, so when it's slaughtered, the meat is not as chalky white as it should be; it has a sort of pinkish hue." Careful not to touch the cod, she takes a closer look. "If we touch the cod, we start the decomposition process. It goes very fast—in only a few days the fish will be ruined. To ensure that it lasts for the twelve days that it's guaranteed to last, you cannot touch it. You must leave it as it has been put down." If the temperature is right, she continues to say, it can take up to five to six days before the fish begins to deteriorate. In Denmark, they had come across cod that for some reason had not been shipped when it should have been. It was fourteen days old, but the visual impression was still very good. "When you do a good job, the fish keeps. That we could find fourteen-day-old fish that looked that good . . . that means that the producer is doing a good job."[20]

Having been in the cod business for quite some time, the controller knows the history of many of the companies that are now packaging cod under the Skrei brand and finds that there are significant differences between them. Companies that have previous experience of fillet production—and therefore have employees who are accustomed to looking for good quality—immediately caught on. The companies that were accustomed to salting or hanging the cod needed correction. "They would be flabbergasted by all the complaints we made. They did not know how to do quality, they did not have employees with that competence," she recalls. "Still, things have improved a lot over the last few years. Things are only getting better and better, largely because the fish branded as Skrei has got such a large place in the market. Earnings from the cod are very good when the quality is good. Also, it is not really a problem if we find errors and report back. Our aim, after all, is not in any way to punish the businesses. This is as much a form of guidance or advice."[21]

With some doubt, the controllers approve the box with the pinkish fish and move on to the next. Here, all the cod are nice and white, but there are traces of blood in the ice—not enough to disqualify it, but enough for a critical remark. The third box is clean, the cod looks exactly like a perfect cod should, but the small amount of ice in the box makes the controllers react. To make sure that the cod is cold enough, they take its temperature, and when this is satisfactory, this box is also approved. Ordered and valued in ways that are rarely, if ever, revealed its consumer, the "most perfect" of cod are ready to go.

PRICES AS OBJECTS AND AGENTS OF VALUATION

This chapter started out with a rather straightforward question: How are fish made into money? Taking the fresh cod industry as our case, we have argued that to answer this question, we must consider commodity prices as being more than something that is made in, by, and for markets. Instead, we have considered the valuation arrangements and tools of valuation involved in prices and pricing. We have further included in our analysis various ways in which cod commodities are imbued with value throughout their production. By doing so, we have shown how these commodities are in fact priced many times over and by way of quite different valuation arrangements and tools of valuation. Prices, we have moreover shown, have a dual character, in that they can be both the objects and agents of valuation. For instance, with the minimum price system, the cod price is the object of a valuation arrangement that came about through political struggle, but is also an agent of valuation, as it affirms the value of the fisheries as key to maintaining the economic vitality of coastal communities. The calculative practice of setting a price, we therefore suggest, can perform more-than-market valuations and be prices not only for markets, but for collective concerns.

Social studies of markets have argued that for a commodity market to be possible a firm relationship must be established between "entities that are able to engage in operations of calculation and judgement" and "entities with *pacified* agency that can be transferred as property" (Çalışkan and Callon 2010, 5; emphasis added). Contrary to this, the chapter has shown that the cod, also after its death, co-modifies the valuations of which it is made part. Whether out at sea, along the landing station production line, or carefully arranged in polystyrene boxes furnished with the Skrei brand label, the cod performs in very specific ways. With its bloodstream vulnerable to stress, its skin tender and easily harmed by sharp objects, its meat soft and sensitive to pressure, it resists being captured and killed in ways that do not comply with its propensities and affordances. It may die the perfect death—at least according to the *Skrei Guide*—but it will still be alive with enzymes and microbes that are busily working to break down its sought-after meat and which must be kept in check by transporting the cod in an unbroken chain of cold storage facilities. Touch the cod and the process speeds up. In more than one sense, then, the cod is a being that must be worked on and worked with—cared for—to be commodified. One wrong move is enough to significantly reduce its quality and the market prices it can achieve. To transform an entity into a commodity, we therefore argue, is not necessarily dependent on its passivity, but can also be achieved by working with its propensities and affordances and through processes of co-modification. Most prominently, its propensity to rot and decay directs the speed of production and trade as well as the material arrangements by which cod commodities travel, from the docks to the market.

The propensities and affordances of the cod complicate the work of upholding its quality and thereby its market value, but are also, as we explored through the work of the Skrei brand, taken as an opportunity to bring it into innovation. Here, we find, the tools of valuation that work toward raising its market value simultaneously work as tools of innovation—tools that draw together and coordinate, that seek to capture the concerns of markets and make them act upon the actions of the industry. At the various sites explored in this chapter, the work performed—negotiating prices, buying, and selling cod; bleeding, cutting, gutting, washing, sorting, packaging, and chilling cod; transporting it for thousands of miles—further tells us that the economization of a cod is not achieved at one point in time. It is neither a straightforward nor a final transformation, but something that must continually be achieved. This in turn shows that not only is the state of cod commodities fragile, but so, too, are the various versions of economization that come about by drawing nature into their fold.

42
36"

7 THE ARCHITECTURE OF MEGA MARKETS

The 2018 China Fisheries & Seafood Expo is about to open, and buyers and sellers of seafood from around the world swarm the front of the expo area. Ten stories tall and made from glass, with a red roof shading the entrance, it welcomes exhibitors and spectators with music and a fountain spurting tall streams of water into the air. Mounted to the glass wall are large banners advertising not only seafood companies, but also seafood nations: "See you at Korea," "Russia," and "Come Visit Canada." This is a place where market and state actors come together, sometimes by way of hybrid agencies and valuation arrangements that build, shift and expand *market architectures* and *market designs*.

Behind the glass facade and past the checkpoints of the registration hall, escalators deliver the visitors to the expo grounds. Wide concrete streets are spread out in a neat system, dividing one expo hall from the next. This year's event is the twenty-third annual China Fisheries & Seafood Expo, and the organizers report that interest is record-high (Redmayne and Sea Fare Expositions 2018). The show spaces in all ten expo halls of the expo center are sold out. Divided over 45,000 square meters of exhibit space, there are over 1,500 companies from fifty-one countries exhibiting their goods to almost 30,000 visitors from 100 countries. No longer just the largest seafood trade show in Asia, the China Fisheries & Seafood Expo is now the largest in the world. The scale of the expo is also a good fit for the stated Norwegian aim of achieving a six-fold growth in the national ocean economy by 2030 (Reinertsen and Asdal 2019)—a "mega" ambition that will be hard to accomplish without taking significant shares of "mega" seafood markets.

Inside the expo halls, an overwhelming diversity of foods from the sea has been put on display: tiny fish and huge fish, round fish and flat fish, bright red and yellow, or pastel green and silvery fish. There is prawn and king prawn; crab and lobster of various sizes, species, and origin; mussels, scallops, and other mollusks; sea cucumbers, sea urchins, and seaweed; roe in all the colors of the rainbow; octopus and oysters. Here, the cod is a foreigner among strangers, the cod, too, looking unfamiliar. Having been put inside a dome-shaped display case, it is here truly domesticated as an object of the marketplace. Neatly adorned on a bed of ice, it is furthermore equipped with a label telling us that it belongs to the export company Lerøy and to the species of *Gadus morhua*, the Atlantic cod. Gutted but with its head still attached to its neck, the cod's dead eyes give the onlookers a glassy stare.

Still, it is perhaps not the cod "in the flesh" that is the most interesting thing here. What makes the 2018 China Fisheries & Seafood Expo an interesting cod site to explore is rather its architecture, an architecture that is closely connected to a market design that both responds to and enacts, in its own ways, the qualities of the mega-sized Chinese markets. This chapter starts out to examine such market architectures and designs by considering the built structures of the 2018 China Fisheries & Seafood Expo and the placement of the "Norway" pavilion within this. Key to understanding these architectures, we argue, is the work of the Norwegian Seafood Council, a valuation arrangement that works by bringing together a wide assortment of tools of valuation. These include what we identify as a form of state-mandated, yet business-driven "seafood diplomacy," as well other tools of valuation such as a consumer survey described as "the largest made for cod ever," a consultancy report forecasting a growth in consumer markets that "made everyone look to China," and various branding and promotional schemes made in response to these. Notably, these are tools of valuation whose capacities are drawn from both the state and market actors, the organizational set-up of the Norwegian Seafood Council being, we will show, key to the hybrid nature of its work. Before going into the details of this state-market valuation arrangement and how it has worked both within and beyond the Qingdao expo grounds, we consider how the social sciences and, most prominently, social studies of markets so far have dealt with market architectures and designs, as well as with the role of the state in market work.

MARKET ARCHITECTURES AND DESIGNS

The social sciences have done much to describe how nations provide a legal and regulatory framework for economic activities, including an extensive body of critical research on the relationship between neoliberalism, markets, and governance (see, e.g., Harvey 2007; Peck 2010; Cahill et al. 2018). What this chapter demonstrates is how the work of government agencies toward the building of markets does not stop at providing legal and regulatory frameworks. Governments also engage in concrete and direct market activities, such as product development, branding, and promotion. They build market architectures and designs. As argued by Neil Fligstein in the book titled, exactly, *The Architecture of Markets*, the relationship between modern state building and modern economy building is key to understanding how economies are built (Fligstein 2001, 8). And so, where Fligstein does much to put the state into the study of market institutions and formations, but without discussing what he takes his title—"the architecture of markets"—to mean more concretely, we demonstrate the coming into being and the workings of such market architectures. In order to do this, we move away from the approach that is largely oriented toward a study of social "structures" and the production of "general rules" (Fligstein 2001, 97) for a more open-ended, empirical examination of economy.

Recent contributions within social studies of markets have drawn attention to how economists have moved away from approaching markets as operating simply through "market mechanisms" and, further, from considering government as the only institution that can correct the possible failures of such a mechanism (Nik-Khah and Mirowski 2019). Previously, the idea was that such "market failures" could be corrected by way of taxes that, for instance, take environmental concerns into account by establishing a tax on pollutants. In this way, so-called externalities—concerns that the market did not "capture"—could be taken into the economic frame and handled (Asdal 1998). By way of economic thinking rooted in neoliberal thought, Edward Nik-Khah and Philip Mirowski (2019) show, this idea was challenged and replaced by the notion that markets needed to be *designed* to work well. Economists, not governments, were moreover put forward as the experts qualified to make such market designs. With this, an understanding emerged where markets were considered both as already very well-equipped and as entities that could be improved by changes to their design.

Such improvements, moreover, were not only considered to manage and solve the best or optimal allocation of resources, but also collective concerns. The climate issue, for instance, was put forward as one that could be managed by markets designed for the buying and selling of carbon quotes.

This shift in economic thinking is captured by the title of the editorial "The Organization of Markets for Collective Concerns and Their Failures," written by Christian Frankel, José Ossandon, and Trine Pallesen (2019) for a special issue of the journal *Economy and Society*. We drew on this work also in chapter 6, though with a somewhat different focus. With the notion of "prices for collective concerns" we demonstrated how the so-called minimum price system generated prices that were intended to serve a collective concern: to make fishing a viable way of living and thereby sustain local coastal communities. We also emphasized that, initially, minimum prices were not organized or modeled on the market. They were the outcome of political struggle and negotiation and relied on the tools and procedures of politics. This valuation arrangement was in no way the outcome of neoliberal reasoning or economics. Instead, it was the result of protracted political struggle, strikes, and the organizational power to make politics adapt and respond to economic demands made by the fishers.

This chapter takes the problematic of market design further, now in a direction closer to the discussions initially introduced by the authors of the special issue of *Economy and Society* (Frankel, Ossandon, and Pallesen 2019). Our analysis furthermore explores the issue of market design as a response to the question of how the architecture of markets comes into place. Yet as alerted to above, the chapter does not consider the architecture of markets as a form of infrastructure enabled by the state. Nor is our focus, as in Nik-Khan and Mirowski's (2019) analysis, the origins of economic thought, from where and whom the ideas of economists came from, or which groups or specific actors introduced and sometimes realized them. Rather, we describe how market architectures and designs come into place and the new valuation arrangements this produces. We consider how the valuation arrangements and tools of valuation used to build them move *across* the public and the private, but also work to transform what the public and the private, a state and a market actor, are. What this chapter follows, therefore, is how market design is made to become a collective concern and, further, is manufactured by way of tools of valuation where the "little tools" of the state, such as its capacity for diplomacy, become major assets for growing a mega market and export-oriented seafood industry. In other words, the state becomes a site *for* the design of markets and for manufacturing this into a collective public concern.

A MARKET ARCHITECTURE OF FRUGAL PROPERTIES

The China Fisheries & Seafood Expo is a place for business-to-business—or so-called B2B—trade. It caters exclusively to buyers and sellers and offers few amenities (say, benches to rest on) to the mere spectator. The expo halls are cold, which is favorable to bringing down the smell of fish but unforgiving to the thinly dressed expo participant. The floors of the exhibition halls are made of concrete, and only in the most extravagant expo areas is there a thin carpet to cushion one's step. Inside the gates of the expo center there are two main "streets," E-street and S-street, which meet in front of the main registration hall. Along each of these lie four exhibition halls, E1–E4 along one street, S1–S4 along the other. Angled so that it intersects with S-street, between halls S2 and S4 is W-street, where halls W1 and W2 are situated. The most central halls—S1, S2, E1, and E2—are clearly the most attractive, as it is here that one finds the largest and most extravagant exhibitions; this is also where there are carpeted floors.

Booming with music, gigantic flat-screen TVs blasting commercials, huge banners hanging from the roof, with little parades of mascots occasionally walking through, these center halls are occupied by the multi-billion-dollar seafood companies, many of them belonging to the Qingdao region. In the marketplace, the architecture of this expo reminds us, not all participants are equal. At the very edge of the expo, at the end of S- and E-streets, are what must be the cheap spots, as the stalls here are small and comparably unimpressive. No business occupies more than one lot, or square, of the hall; there are no flat screens or loudspeakers; and business seems remarkably slower than in the central halls. In W-street, fish are few and far between, as these are the halls of the hardware, showing engines and production line systems, the buzz and the business of these W1 and W2 halls being far less busy than those of the central seafood halls.

Compared to department stores or shopping malls, which are carefully designed to entice visitors to both extend and repeat their shopping experience, the Qingdao International Expo Center comes across as a rather cold and hard marketplace, ripped down to the minimum. There are food courts, but they are not inviting; there are public restrooms, but they do not evoke a sense of comfort. This is a marketplace of rather frugal architectural properties, and it is a strictly bounded place. It has gates and fences, and to get in you need to pay a rather high entrance fee. And still, it is also very much a connected site as it brings together, from around the world, buyers and sellers under the roofs of its halls. This bringing-together of market actors does not come about easily, however. In the case of the cod, it depends upon the Norwegian Seafood Council, a valuation arrangement that works by use of several tools of valuation, for example by what we suggest thinking of as a form of "seafood diplomacy"—diplomat bodies and state agencies working on behalf of the seafood nation and its seafood companies to increase export volumes and values. As signaled by the banners hanging from the expo facade, both seafood companies and seafood nations are at work here. The architecture of this marketplace is tightly woven together with state machineries, the expo and its rich assortment of seafood being inextricably political-economic. This is also the case of the "Norway" pavilion, where the political-economic materializes under the slogan "origin matters."

"ORIGIN MATTERS"

When entering the 2018 China Fisheries & Seafood Expo, visitors receive a tote bag containing a pile of printed advertisements and a show directory, a thick book with maps of the expo grounds and an overview of the exhibitors. The maps show the floor plan of each hall and where to find the exhibitors, while the overview details the names of the exhibiting companies, their categories of trade, and where on the expo grounds they are to be found. The designation of the Norwegian companies shows that most of them are not at the very center of the expo grounds, but right next to it, in Hall E2. Like E1, this hall is reserved for foreign seafood exporters. Individual companies occupy some of the lots, but it is predominantly organized through so-called pavilions, which are occupied by the companies of one nation. The "Norway" pavilion is situated toward the middle of the hall and is among the largest, but it is still considerably smaller than the pavilions branded "USA" and "Russia."

Lodged in between the "USA" and "Korea" pavilions, the "Norway" pavilion is demarcated by large posters hanging from the ceiling. On the ground level it is compartmentalized into different areas belonging to different

FIGURE 7.1
"Norway" pavilion, at the 2018 China Fisheries & Seafood Expo. Photograph by Tone Huse, November 7, 2018.

companies. Most of the companies are from the salmon farming industry, but the fisheries are also well represented. One of the companies, Lerøy, has exhibited what looks like the entire Norwegian seafood menu, including the cod resting inside the dome-shaped display case. Others have opted to display their goods not in the flesh, but by way of videos and graphic renderings of the different species that they are there to promote. Thin walls separate one compartment of the pavilion from the next and company logos are inscribed upon the walls, demarcating which company the different areas belong to: "Lerøy," "Coast," "Domstein," "Seaborn," "Nordlaks," "Norway Royal Salmon," "Norwegian Seafood," "Ocean Supreme," "Sekkingstad," "Polar Seafrozen," "Global Fish," "Bravo Seafood," "Kuehne + Nagel," "Pelagia," and "Ocean Quality." In between the many logos, potential Chinese buyers mingle with Norwegians from the different export businesses and their Chinese interpreters. The pavilion posters feature photographs of Norwegian ocean landscapes. The photographs are dominated by the color of blue, its different nuances reflected in images of the ocean, skies, and snow-clad mountains. Some of the posters are inscribed with "Norway," others with the slogan "Origin matters," printed in both English and Mandarin.

The "Origin matters" slogan has not been made specifically for the expo but is used globally by the Norwegian Seafood Council as part of a promotion strategy of associating seafood from Norway as having an origin in cold, clean waters. Here, then, we encounter an "origin story" (Haraway 1989, 1997) that is similar to, but also different from that explored in chapter 5. The "Origin matters" slogan may well allude to the "proud" traditions or current capacities of the Norwegian seafood industry, but it is the place and the natural environment from which the seafood originates that are being emphasized. In the context of this 2018 event, with frozen seafood being displayed in abundance, there is also a certain irony to the use of this slogan. For at the time, only a year had passed since Norway was itself let out of what journalists termed the "Chinese freezer," a diplomatic chill spot it was promptly put in when the Norwegian Nobel Committee awarded the 2010 Nobel Peace Prize to the Chinese dissident Liu Xiaobo. The Chinese government blamed the Norwegian government for this, and relations soured despite the Norwegian government's efforts to explain that the Nobel Committee is an independent body. In 2017, diplomatic connections were reopened, which also created expectations that the Chinese embargo on certain Norwegian goods—the salmon being among the most important—would soon be lifted.

Not only fish, however, but also high-ranking officials were now set up to pass the border: in the time between the 2017 thaw and the 2018 China Fisheries & Seafood Expo, official visits to China were made by King Harald V and Queen Sonja, who are the formal heads of state in Norway, and by the Norwegian Prime Minister, the Minister of Foreign Affairs, the Minister of Finance, the Minister of Culture, the Minister of Petroleum and Energy, the Minister of Climate and Environment, the Minister of Health and Care Services, the Minister of Research and Higher Education, the Minister of Trade and Industry, and the Minister of Fisheries.[1] The Minister of Trade and Commerce visited twice, as did the Minister of Fisheries. "I love China!" he proclaimed from a stage in Beijing (Stanghelle 2017).

In tow during the Minister of Trade and Commerce and the Minister of Fisheries' visits were 348 "accompanying" businesses, institutions, and organizations. Of these, sixty-two can be directly linked to the seafood industry. It was, in other words, beyond doubt that in selling fish to China, "origin matters" not only with respect to the waters from which the seafood comes, but also with respect to its nationality. That the Norwegian government had been quite eager to improve the status of this nationality furthermore came to view in how it did very little to press China on sensitive political issues, such as human rights violations. This omission was repeatedly criticized in Norwegian public debate, and by the Chinese dissident Hu Jia. "Who is it that comes to visit?" he asked rhetorically (Svaar, Hotvedt, and Hirsti 2017). "Is it Norway's Prime Minister or is it a fish seller?"

Jia's question is straightforward, but the distinction between the public and the private can be quite hard to make. In many cases, government organizations and officials represent the economic interests of the nation, a role that is at odds with being "purely" political. More so, the market ambitions that were being pursued at the 2018 China Fisheries & Seafood Expo—by way of the Norwegian Seafood Council—is of a thoroughly hybrid nature. This an ambiguous status to the market architectures and designs that it is both a part of and seeks to build. The Norwegian Seafood Council was in fact the main designer of the "Norway" pavilion and had, in preparation for the expo, assisted the participating companies in various ways. It gave advice on how to acquire an entry visa to China, on which smartphone app to download to keep in touch with new Chinese contacts, and it arranged various pre-expo warm-up events. If we are to consider the "Norway" pavilion as a particular marketplace within the 2018 China Fisheries & Seafood Expo and, further, assess which type of market architecture and design is here being made, we must therefore take a closer look at the Norwegian Seafood Council and how it works, precisely, as a hybrid agency and valuation arrangement.

A HYBRID VALUATION ARRANGEMENT

Formally, the Norwegian Seafood Council came into being by the passing of the 1990 Fishery Export Law.[2] The law was the outcome of a decade-long legal revision process and led to the dismantling of a complex export control system organized through a myriad of so-called export associations and mandated sales organizations (Holm 1995). For cod, there were eight such mandated sales organizations, which by 1991 were replaced by the one export council, today known as the Norwegian Seafood Council.[3] The ownership of the Norwegian Seafood Council was placed within the Ministry of Fisheries,[4] but it was to be funded, exclusively, by obligatory fees paid by the seafood export businesses. It was further decided that the head office of the Norwegian Seafood Council was to be in the North-Norwegian city of Tromsø. Its primary tasks would be to grant export licenses to Norwegian seafood companies and to oversee that the quality of the seafood exported from Norway was satisfactory, the fear being that the export of low-quality seafood could damage the reputation of the entire industry.

In the paper trail of policy documents leading up to the Fishery Export Law of 1990, as well as in the minutes from the parliamentary debate on its passing, this new way of organizing is repeatedly characterized as a "liberalization" and "deregulation" of fish exports.[5] According to fishery scientist Petter Holm (1995, 414), the dismantling of the system of multiple mandated sales organizations allowed exporters to escape from a "fine-meshed system of regulation" where the sales organizations acted as both interest organizations and cartels. Still, when reading more closely the policy documents and the expectations they chart for the Norwegian Seafood Council, it becomes clear that while this new way of organizing the licensing of seafood exports was far less bureaucratic than the system it replaced, it also entailed the placement of a new type of economic work within and as part of the state apparatus. Most importantly, in specifying the tasks of the new export council, the Ministry of Fisheries established that it was to be market-oriented and lead the way in executing "common domestic and foreign market measures and other export promotion work."[6] Still, and due to how the Norwegian Seafood Council is funded by a fee paid by the export businesses, the priorities of this work should be set not by the Ministry of Fisheries but by the industry itself, so as to ensure that this taxation is always to their benefit. As put by one of the market analysts employed at the head office in Tromsø: "If we don't create value for the industry in some way or other, we do not have a purpose."[7]

The Norwegian Seafood Council's industry affinity is reflected in its formal organization, where industry representatives dominate both the board of directors and the so-called market groups that advise the council's different departments. The board of directors decides how the income from the export taxation is to be spent, and currently about a third is set aside for shared tasks, such as administration, market research, and the running of foreign offices.[8] The remaining two-thirds are used for species-specific work, and this is indeed species-specific: if income from cod exports amounts to 10 percent of the Norwegian Seafood Council's income, this "cod-income" will cover 10 percent of the shared expenses. The remaining 90 percent of the income from cod exports will be spent on cod-specific promotions, such as cod market surveys or cod promotion events. The species-specific financial structure is further reflected in the Norwegian Seafood Council's organizational structure, where its market groups are organized around "white fish" (which includes fresh and frozen cod), "conventional" (which includes dried and salted cod), "salmon and trout," "pelagic fish," and "shrimp and shellfish."[9] For each of these market groups there is a board where a majority of representatives are from the industry, and it is this board that makes all the strategic decisions regarding future market work. It was therefore the white fish market group that in 2014 decided that the Norwegian Seafood Council should, first, pursue a strategy of building new markets for cod and, second, do so by exploring opportunities for building a domestic market for cod in China.

TEAM NORWAY AND THE GENERIC ART OF SEAFOOD DIPLOMACY

When considering the Norwegian Seafood Council as a valuation arrangement, an important aspect to note is how it is tasked to represent and enhance market values for the entire seafood industry, and not particular companies—a premise that is key to the status of its so-called delegates, who are employees stationed in foreign offices. In addition to the Tromsø head office, which employs about sixty-five people, about fifteen Norwegian Seafood Council offices have been established in what is considered as strategically important markets around the world:[10] Brazil, China, France, Germany, Italy, Japan, Poland, Portugal, South Korea, Spain, Sweden, Thailand, the United Kingdom, the United States, and West and Central Africa. The delegates at these sites are responsible for carrying out the Norwegian Seafood Council's work and do so by being part of the Norwegian diplomatic corps. Following the Foreign Service Act,[11] the delegates have the diplomatic status of so-called other delegates, which is reserved for persons employed by independent legal entities and who often act as vice consuls of either a Norwegian embassy or a consulate. When feasible, the Norwegian Seafood Council's delegates share offices with the embassy or a consulate, and are then part of the so-called Team Norway, a network-based collaboration between different public and private actors working internationally to advance Norwegian business interests.[12] In compliance with the Vienna Convention on Diplomatic Relations, the Norwegian Seafood Council cannot engage in direct commercial activity[13]—buying and selling seafood—and must limit itself to advancing *general* Norwegian interests. These activities are *not* to benefit select businesses, and any marketing activity carried out by the Norwegian Seafood Council must therefore be generic: its number-one brand is "Seafood from Norway"; its number-one slogan is "Origin matters." It is further committed to promoting the values of "quality," "sustainability," and "health" as specific to Norwegian seafood products. As with the Skrei quality brand discussed in chapter 6, the Norwegian Seafood Council polices its image carefully. It makes sure that the image is not used to market non-Norwegian seafood, and has a stand-by division that, in the case of negative attention in foreign media, can mobilize and work on improving public relations. Additionally, the delegates stationed abroad are very important to the various campaigns initiated by the different market groups and are thereby responsible for carrying out a wide assortment of market work.

In sum, the Norwegian Seafood Council is a state-initiated valuation arrangement that draws on the apparatus of the state to build market architectures and designs and perform its market work. Most notably, this comes to view in how it is part of the Norwegian diplomatic corps. This equips the Norwegian Seafood Council with the capacity to bolster the seafood industry's position as an export economy at large, while aiding the individual companies in pursuing increased export volumes and values. In doing this, the Norwegian Seafood Council is actively designing markets, while simultaneously drawing on the state and its capacities to act "diplomatically." It acts as a valuation arrangement that works toward enhancing market values by letting market work extend to equipping seafood with diplomatic associations and thereby the ability to move more smoothly. It is from this, then, that our notion of seafood diplomacy comes from.

The notion of diplomacy should not, however, be taken to mean that the hybrid operations of the Norwegian Seafood Council are tension-free, not even "internally," among the very companies that are supposed to benefit from its work. Indeed, the role of the Norwegian Seafood Council is regularly debated and criticized, something that also came to be expressed at the 2018 China Fisheries & Seafood Expo. For while most of the Norwegian export companies gathered under the banner of the Norwegian Seafood Council, Marine Harvest, one of the largest producers of salmon and trout in the world, did not. Instead, it set up a separate exhibition right next to the "Norway" pavilion, the area it occupied being larger and more extravagant than any other area occupied by an individual Norwegian company.

By being one of the major Norwegian export companies, Marine Harvest funds a major share of the Norwegian Seafood Council's "salmon and trout" work but has objected to this on several occasions. Instead of funding what the company sees as a disproportionately large share of marketing that benefits the entire industry, and therefore also its competitors, Marine Harvest would prefer to spend the funds on marketing that benefits its own products (Nodland 2016). As explained by one of the company's representatives in the newspaper *Nordlys*, the Norwegian Seafood Council occupies a position that the company wants to take for itself: "When they pour out ads, they take a position in the head of the consumer, and then they supplant our position. The Norwegian Seafood Council promotes 'Norway' as a brand. When, for instance, Lerøy wants to get into a market, they get competition from 'Norway'; all the while they themselves are part of financing 'Norway'" (Alexandersen 2014). Also, as Marine Harvest has production facilities not only in Norway but also in Canada, Chile, the Faroe Islands, Ireland, and Scotland, not all of its products will in fact benefit from the "origin matters" strategy. Consequently, the valuation arrangement of the Norwegian Seafood Council, with its focus on the value of Norwegian products, is not considered as serving all market actors equally well. This internal competition can be also read as a sign of the growth that is expected from the Chinese seafood market in the years to come, and how Marine Harvest and other large seafood actors are preparing and situating themselves to be able to profit from it. As we now move to exactly this—the prospect of China as a future mega market—we address how the enactment of such future markets can be considered as a way of creating a market architecture and design.

"CHINA'S NEXT CHAPTER": A MEGA CONSUMER ON THE RISE

There were mainly two reasons why China was selected as a new market for Norwegian cod.[14] The first has to do with the supply of fish. Cod quotas had for quite some time been relatively high and concerns were being voiced that existing markets were soon to be saturated. The second reason is related to the idea of increasing market robustness for the white fish trade by diversifying its market positions through emerging consumer markets. By acquiring positions in more markets, one assumed, the export companies would become less vulnerable to local variations in, for instance, fluctuating purchasing power or, as was the case with China, being put in the diplomatic freezer. In 2014, the white fish market group of the Norwegian Seafood Council therefore decided to pursue what they called a "New Markets Strategy."[15] Several countries were surveyed as potential targets, and China was selected as the first trial market for the new strategy, a decision that was largely influenced by the forecasting report "China's Next Chapter," published by the international consultancy McKinsey & Company in 2013.[16]

"China's Next Chapter" starts out by stating that the Chinese economy is about to undergo fundamental change and needs "rebalancing." To maintain economic growth, the report argues, China must reduce its reliance on investments in a production-oriented economy and instead increase its domestic consumption (McKinsey & Company 2013, 35–36). Following up on this, the report presents an analysis of what it calls "the explosive growth of China's new middle class" (McKinsey & Company 2013, 54), citing forecasts produced by McKinsey & Company that by 2022 more than 75 percent of Chinese consumers will earn an annual income between 60,000 and 229,000 renminbi, equivalent to about $9,000 and $34,000. In contrast, only 4 percent of the country's population belonged to this income class in 2000; by 2012 this share had grown to 68 percent. Importantly, the report continues, the consumption of this new middle class will be dominated by "the 'upper' cut" consisting of "sophisticated and seasoned shoppers" who have not only the means but also the willingness to pay "a premium for quality" (McKinsey & Company 2013, 55). Notably, divides in consumption are expected to follow along generational lines, as a new culture of consumption is seen as being on the rise within the so-called Generation 2 (G2) of China's new middle class:

> These G2 consumers today are typically teenagers and people in their early 20s, born after the mid-1980s and raised in a period of relative abundance. Their parents, who lived through years of shortage, focused primarily on building economic security. But many G2 consumers were born after Deng Xiaoping's visit to the southern region—the beginning of a new era of economic reform and of China's opening up to the world. They are confident, independent minded, and determined to display that independence through their consumption. . . . McKinsey research has shown that this generation of Chinese consumers is the most Westernized to date. . . . These consumers seek emotional satisfaction through better taste or higher status, are loyal to the brands they trust, and prefer niche over mass brands. (McKinsey & Company 2013, 57)

As stated by the Norwegian Seafood Council's delegate to China at the time, the effect of the McKinsey & Company report was to make "everyone" look to China. "It was one of the first reports that considered the Chinese consumer," he explains. "Before that most had considered China as a country to produce things in, that we could buy things from. It was quite novel, that one considered China as a consumer." When in 2014 the Norwegian Seafood Council decided to pursue a strategy of exploring new markets for cod, China was therefore selected as a test market. Since then, the domestic consumer market of China has indeed risen to become one of the world's most coveted markets. Considered to be a still-growing mega market—and a seafood-loving one at that—it is highly attractive to Norwegian seafood exporters. But how, exactly, does one go about taking a share of such a market? Which tools to use? In other words, how to revalue China from being a producer to becoming a consumer, and how to design a market toward that end?

Unlike salmon, which was embargoed during the seven-year-long diplomatic freeze between Norway and China, the import of cod products went on unhindered. A reason for this differentiation was never given, but one could speculate that it was connected to the different market positions of the two species. The salmon trade generates a much higher profit than the cod, which means that embargoing it would harm the Norwegian economy more. Salmon is moreover mainly imported for consumption at sushi restaurants, while the cod is used by the Chinese seafood processing industry. In fact, roughly 50 percent of the frozen cod exported from Norway is sold to Chinese importers.[17] This frozen cod is largely captured by the oceangoing fleet, which consists of trawlers, long-liners, and Danish seine boats. The fish is frozen upon capture and stored in large onboard freezers, before it is landed at so-called neutral freezer hotels—that is, facilities that consist of huge freezer halls with storage for thousands of tons of frozen fish.[18]

Like the cod that is landed fresh (see chapter 6), there is a minimum price system in place for cod that is landed frozen. If the owner of the fish has not already made an agreement with a fish trader, the fish is sold by way of a digital auction that takes place on the website of the Norwegian Fishermen's Sales Organization (see chapter 6). The fish traders buy and sell cod with low margins, earning only a few cents per kilogram of fish bought and sold. They therefore rely on trading in large volumes, which makes the large Chinese factories attractive buyers. It is not unusual, however, that ownership of the fish changes hands several times before it is shipped out. For contrary to fresh fish, which must be rushed to markets, the frozen fish can last longer in storage, giving it a commodity status closer to more inert objects. This opens up the frozen cod trade for speculation and drives prices up. Consequently, while frozen cod is generally considered to be of a lower quality than fresh cod, it is nonetheless sold at a higher price per kilogram. This price difference, however, could also be related to how it more easily enters other forms of production than the fresh cod sold at restaurants and supermarkets. For once the cod arrives in China as a raw material, the frozen cod has traditionally been processed and re-exported. In fact, almost all of the seafood that China imports, about 99 percent, is re-exported to markets abroad, the EU and the United States being large importers.[19] If you are a European or North American consumer of products made of cod captured in Norway—for example, portion-size fillets or battered fish fingers—there is a good chance that this product is made *not* in Norway, but in China (Lysvold 2019).

The position of the cod as a raw material in the Chinese seafood industry could well have protected it from the import restrictions that affected the salmon industry. However, from the point of view of the New Markets Strategy launched in 2014, it was also a challenge. How was one to go about shifting the cod's status from being a raw material of the export industries to becoming a preferred item on Chinese dinner tables? How, in other words, to change its commodity propensities? As we will see below, this was approached through several of the tools of valuation that the Norwegian Seafood Council had at its disposal. This centered, first, on identifying a cod desirable to Chinese consumers and, second, on lending the cod and its producers the authority of the state, equipping it with diplomatic associations on top of its commodity value. Throughout this work, the "Origin matters" slogan and "Norway" branding—were put to work, underscoring how nationhood and economy can come together in the practices of building a market architecture and design.

VALUING WITH A SURVEY

How to build a market architecture and design that caters to a market expected to grow huge and assume "mega" qualities when not knowing much about the terrain on which the market is to be built? Not unlike how companies operate, the Norwegian Seafood Council started off this work by conducting a market analysis.[20] Over the course of six months, thirty-four focus group interviews were held, and 500 restaurants and 250 shops were surveyed. The Norwegian Seafood Council's delegate to China led the work, describing it as "probably the largest market survey done for cod anytime and anywhere"[21]—a mega survey for a mega market—and a tool of valuation that was to revalue what kind of market was about to emerge in China and how to cater to it.

Surveys are performative, John Law (2009) argues in the article "Seeing like a Survey." They *do* things to the objects and subjects they are targeting. In the case studied by Law, a Eurobarometer survey of people's attitudes to animal welfare, the survey enacted a specific form of consumer—a neoliberal consumer with the capacity to choose. Likewise, the McKinsey & Company report above enacts "the Chinese" as, exactly, a consumer. It thereby partakes in a quite strong move, away from that of capturing China as a mega producer, and toward casting it as a mega consumer. Yet, reports and surveys like the ones in question here do more. They do valuation work and act as tools of valuation. Considering how valuations are performed we can furthermore detect an interesting difference between the report of McKinsey & Company and the ensuing Norwegian Seafood Council survey. The McKinsey & Company report seems to work very much along the lines of what can be described as a standard Western approach in and for business. It addresses how Chinese manufacturing "moves up the value chain," while also underlining how the Chinese middle-class consumer is up for "higher-value products" (McKinsey & Company 2013, 1). The report thereby enacts a specific value ordering, where value is considered within a more or less given system and as being situated lower or higher up a "value chain," these positions giving value to "lower-" and "higher-value" products.

The market survey of the Norwegian Seafood Council complicates this value ordering by bringing in another consumer altogether—a consumer who values differently. For what they learned from the survey, the Norwegian Seafood Council delegate explains, is that Chinese consumers, however "Westernized," are also quite different from those in Norway:

> Take chicken in China. I have a dog, a small poodle. His snack is chicken breast. So the part of the chicken that is priced the highest and valued the most in the West and in Norway, that is the one that is priced the lowest in China. What the Chinese like is the wing; chicken feet are very popular; the throat; chicken with bone. If you take the skin off, and the bones out, what you are left with is a shoe sole. The same goes for fish. Therefore, my most important message to the cod industry, since we got the results of the market research that we did in 2014, is that the largest volume will be cod cutlets. The cutlet format had few negatives in China. That it comes with the skin on is no problem. Or, put differently, fish without the skin is seen as ridiculous in China. To take the skin off is to ruin the fish. So why take the skin off—they love it. It has a lot of nutrients, it is where the collagen is, which makes *our* skin nice. And the skin keeps the fish fresh. Therefore, fish should have skin. And bone. Bones also had many positive associations. They find that when they prepare the cod, steam it or something, it keeps fresher if it has the bone attached. And like the skin there are a lot of nutrients in the bone. So, there are no negatives with cutlets. And it is more efficient to produce, so you get a lower price.[22]

What emerged from the Norwegian Seafood Council's survey, then, was a wholly different mode of value ordering. Here, values are not *given* in a predefined value chain system, nor are they simply values moving smoothly along axes of "lower-" versus "higher-value" products. Rather, what the survey performed was a revaluing of, first, the seafood commodity in question and, second, a consumer who was valuing cod for completely different reasons and in completely different formats than what the Western taste and gaze expected. Interestingly, then, when the Norwegian Seafood Council's market survey came to "see" its market differently, this was due to its ways of drawing the consumer and the cod together, enacting a consumer–commodity assembly and performing a co-modification of consumers and the products presented to them. What we see here, then, are the contours of a valuation arrangement in formation, an arrangement, moreover, that would be decisive to how the Norwegian Seafood Council approached the challenge of creating a market architecture and design that could align the cod and the Chinese consumer.

By way of the market survey, four other revaluations were furthermore added to this emergent valuation arrangement. First, a revaluation of cod, which was considered not as an input to factory productions of seafood products, nor a luxury fish like salmon, but rather an everyday fish to be prepared and consumed at home. The pricing of cod commodities, the Norwegian Seafood Council therefore concluded, had to be done accordingly. Second, the taste of cod was of a different value than assumed, as it was taken to not have much taste, and instead be a carrier of taste enhancers. This meant that it could be promoted as a meat that blends quite easily with the many and quite different kitchen traditions of China. Third, the cod's origin was taken to be of much higher value than so far imagined, as the market survey showed that perhaps the largest product advantage of the cod did not have anything to do with the fish itself, but rather with its Norwegian origin. Origin *did* matter and was highly valued, or so it seemed. The Chinese, the market survey stated, associated Norwegian seafood with the deep sea, a place where the fish would need to struggle for its survival and that therefore would deliver strong and healthy fish. Future promotions were consequently to build on these positive associations of the cod's origin as "safe and healthy," while being a "do-it-yourself" fish that was both "delicious and easy to prepare." The fourth revaluation brought on by the survey, however, was more challenging. The Norwegian cod, it turned out, did not have a Chinese name. To inscribe and imprint on it specific qualities and associations—quite literally, branding the cod—gave little meaning if not having a name meant that it also did not exist, at least not as a discernible market commodity. Rather, the cod was referred to as *shuǐ zú* (水族), a collective term for aquatic animals which was problematic for different reasons. On the one hand, *shuǐ zú* brought forth negative associations with previous food scandals caused by cheap *shuǐ zú*; on the other hand, it could be associated with very exclusive products that cost up to NOK 8,000 per kilogram and which are also described as *shuǐ zú*. The Norwegian Seafood Council therefore opted to convince the Chinese factories that were to produce cod cutlets for the domestic market that the cod should be branded as "*běi jí shuǐ zú*—Norwegian Arctic Cod," *běi jí* meaning "from the Arctic Pole." By way of the market survey, a revaluation process was being shaped, where, in our own words, the new Chinese consumer was to be carefully co-modified with the invention of a whole new consumer–commodity assembly.

MEGA MARKETS FOR A "MEGA" ECONOMY

A tool of valuation, the Norwegian Seafood Council's market survey became key to the ordering of value in the cod market. Through this, it paved the way for what we can recognize as a valuation arrangement that is still under formation but carries the potential of carving out a market architecture and design tailored to the *běi jí shuǐ zú*. Still, this does not capture the valuation arrangement of seafood diplomacy in its entirety. The Norwegian Seafood Council's survey worked to enact a specific version of the consumer and what we capture as a consumer–commodity assembly, but equally important is how the Seafood Council's delegates make themselves available to the companies, thereby adding to the cod a form of status value. "I have no fish to sell," the delegate explains, "but the status that being a diplomat gives, and to be associated with the public authorities of Norway, that is the main value. The status it gives them, it cannot be measured in money."[23] Not aimed directly at consumer markets, this is about enhancing the value that the cod holds in the Chinese seafood industry, making it more attractive to develop and launch as a commodity by adding a particular form of social value to the commodity. Following this, the main activity of delegates is to participate in events initiated by the Chinese industry, which in turn leads to a lot of travel. "It is not unusual that I spend two days getting somewhere and back," the delegate explains, "and the specific thing I do is to make a ten- to fifteen-minute speech. Perhaps participate at a dinner or cut a ribbon. In a Norwegian context, it sounds ridiculous, but the status of a diplomat in China is very much higher than in Norway. It has enormous value for these actors, which in turn ties them very closely to the Norwegian origin and to suppliers from Norway. I show that I support them; in return I get their loyalty."[24]

The diplomacy of communicating across cultural barriers also entails working toward the Norwegian exporters, so that they can better understand what types of fish the Chinese consumer would want to buy. The desired product format of cutlets, for instance, was not something they could provide, as most Norwegian companies were focused on Western markets and their preference for either whole head-cut fish or boneless and skinless cuts, the loin being valued the highest. The Chinese demand for skin and bone, for cutlets and cod heads, was hard to meet. It therefore fell to the Chinese factories to process the cod intended for the new home market, but they also had difficulty adopting the cod to Chinese preferences. For as much as the fish captured by the deep-sea fleet is frozen in blocks—the fish squashed together in neatly formed, rectangular blocks of about twenty kilograms apiece—they would have to defrost the blocks to separate the fish and cut it into portion pieces. Freezing it again led to a lowered quality, and so, as the domestic market for cod began to pick up, the factories began to demand single frozen fish. Again, the demand was difficult to meet, as the entire production line onboard the fishing vessels would need to be rebuilt. "There are language barriers, cultural barriers," the delegate concludes. "The Norwegians don't really get why the Chinese are so difficult. Why can't they just buy that block of frozen cod? Still, there are some that are beginning to adapt. Things will slowly begin to turn in the right direction."[25]

One of the main events organized by the Norwegian Seafood Council in conjunction with the 2018 China Fisheries & Seafood Expo was a seminar that gathered Norwegian export businesses to a crash course in Chinese markets and trade. Branded as a "Norway–China Seafood Summit Seminar," this event took place not within the stripped-down properties of the expo grounds, but in the luxurious surroundings of the five-star Hyatt Regency hotel in Qingdao. The take-home message delivered by the Norwegian Seafood Council's market analyst, flown in from the Tromsø headquarters for the occasion, was also fitting of the surroundings. Giving a presentation titled "Big Picture, Demand and Trade Flow," the analyst gave a presentation that could not be misunderstood: the consumer power of the Chinese domestic market is growing, its consumption of high-end seafood is growing, and there should therefore be ample opportunity for Norwegian seafood exporters to gain market shares. Pulling up a slide showing a pie chart on the giant screens of the seminar room—the slice of "Asia Pacific" colored in dark blue—the analyst furthermore introduced a forecast produced by the World Bank (see figure 7.2). By 2030, she told the audience, Asian consumers are expected to make up 66 percent of the world's middle class. The consumers of Europe and North America will no longer dominate markets.

As presented at the Hyatt Regency, the numbers on the analyst's pie chart come across as quite neutral descriptions of where to concentrate future marketing efforts. When looked at in view of consumer history, however, another, rather striking feature can be added to them. For if these World Bank forecasts come true, Asian markets will no longer represent

The market potential is "unlimited"

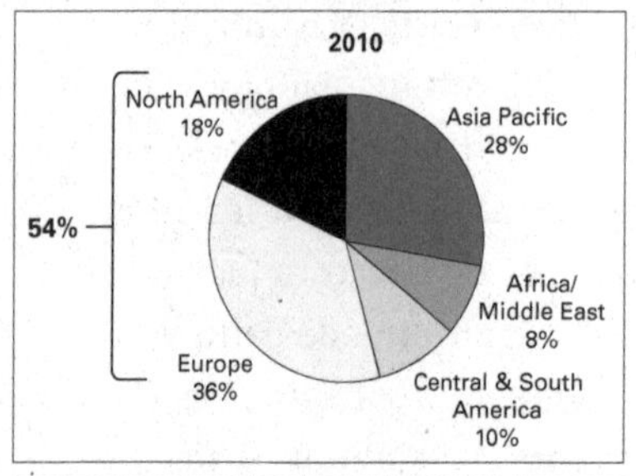

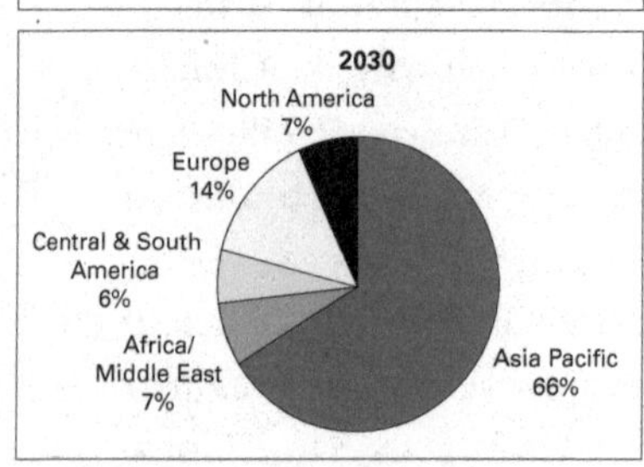

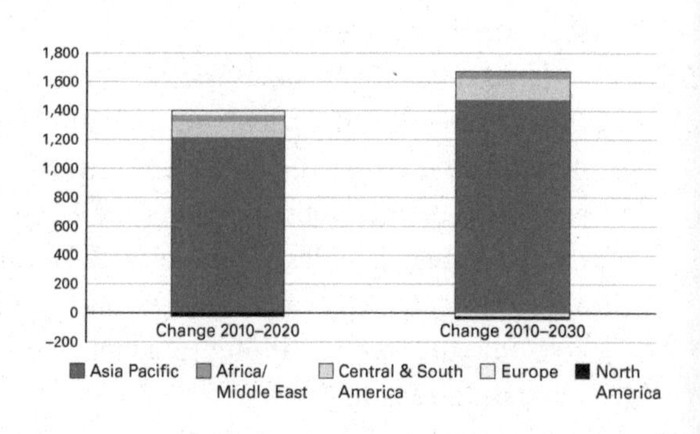

FIGURE 7.2
Norwegian Seafood Council presentation at Norway–China Seafood Summit Seminar, Morning Session, November 6, 2018.

"just" growth opportunities. They will also become the hegemonic driver in defining consumer preferences and put an end to a Western dominance as old as consumerism itself. Such concerns, however, were not present in Qingdao. Here, markets were markets, and as the Norwegian Seafood Council analyst assured the audience: "There will be more Chinese seafood mouths to feed." Stressing the need for seafood exporters to be proactive and get ahead, her slide said it all: "The market potential is 'unlimited.'"

BUILDING MARKETS, EXTENDING THE STATE

In this chapter, we have described the meeting between, on the one hand, the ambitions of Norway in growing its seafood economy and, on the other, the expectations of China becoming a mega consumer market able to absorb much of this growth. The 2018 China Fisheries & Seafood Expo can in that respect be considered as one of the sites where the great economization of the ocean unfolds—not in the form of oil wells piercing the seabed or aquaculture installations crowding coastal waters, but through efforts to create a market for growth to come. And yet, as we have emphasized throughout this book, to understand how such "great" change comes about, we must also consider the "little tools" involved in this. How are these tools and the apparatuses of which they are a part ordering and reordering value, enhancing, assessing, and enacting the value of goods and commodities? The chapter has examined this by focusing on market architectures and designs, approaching markets not in the abstract, but as situated and tangible, material, and semiotic. We have moreover focused on the work of one organization in particular—the Norwegian Seafood Council—which we identify as a valuation arrangement. Characterized by its distinctly hybrid properties and tools of valuation, the Norwegian Seafood Council operates by drawing on the capacities of both market and state. Following up on how chapter 6 discussed what we there call "prices for collective concerns," we have explored in this chapter what social studies of markets have identified as "markets for collective concerns." Whereas studies have connected this notion to a neoliberal mode of rectifying market failure, "market design" being the key tool for doing so, we have in this chapter sought to pin down what such practices of market design can be in the context of a market work that also involves the agencies and operations of the state. And furthermore, that do not have market failure, but market growth as their main concern and aim. In doing so, we have taken on the challenge issued by Fligstein (2001) to examine how modern economy building is also about modern state building, but by way of a much more practice-oriented and, indeed, literal, approach to what Fligstein identifies but does not describe—the architecture of markets. Through this, we have emphasized how market architectures and designs come about not simply by the agencies of market actors or interests, or by way of transforming the state in the image of the market. They are also made by putting to work tools of valuation that can work across state and market agencies, and across the boundaries usually seen to separate the state and the private.

By focusing on markets as a physical, situated reality in which specific valuation arrangements and tools of valuation are put to work—the Norwegian Seafood Council and its seafood diplomacy, market surveys, events, and activities—the chapter has identified how the state extends its capacities and lends its authority to such arrangements and tools. These are tools of valuation that, characteristically, are market tools and state instruments simultaneously. And while they blur the boundaries between these spheres, they do not simply reduce or weaken the state by making it more "market-like" or market-oriented. The valuation arrangements and tools of valuation that this chapter examines build markets, *and* they build the state. In this way, we find, the notion of markets for collective concerns should also be taken as an opening to exploring how state involvement in market architectures and designs can work to grow the state, giving it a corpus and reach beyond the political and deep into the marketing work commonly seen as private. As so precisely captured by the Chinese dissident Hu Jia, in asking whether it was Norway's Prime Minister or a fish seller who came to visit (Svaar, Hotvedt, and Hirsti 2017), the very hybridity of the Norwegian Seafood Council lends itself not only to marketing purposes but also to the purposes of a state eagerly seeking to repair its relations with the Chinese superpower and thereby grasp a share in what is forecasted as "unlimited" market growth. This suggests that future studies of both states and markets should be open to this interchangeability of growing markets and growing the state, including how the "long arms" of the state can extend themselves and reach further by lending themselves to and drawing on hybrid valuation arrangements and tools of valuation. As the many pavilions at the 2018 China Fisheries & Seafood Expo show, not only Norway, but other states also—most prominently Russia, the United States, Canada, and Korea—were here in play, suggesting that this is not only a Norwegian phenomenon. Whether these nations' seafood diplomacy is like that of Norway or not remains to be examined, but both similarities and differences are likely to exist.

56.
1/3

8 GOOD NATURE, GOOD ECONOMY

"Animal species have come and gone but the cod has swum its own course, humanity is just a short span in its life," writes the Icelandic novelist Jón Kalman Stefánsson (2010, 65). Describing the cod as a fish whose insatiability and greed is surpassed only by that of humans, he takes his readers to the very depths at which the cod swims:

> The cod is yellow and enjoys swimming, is always on the lookout for food, very little that is remarkable occurs in its life and a line that sweeps down with bait on a hook is considered great news, it is a huge event. What's this? the cod ask each other, finally something new, says one, and bites immediately, and then all of the others hurry to bite as well, because none wish to stand apart, it's excellent to hang around here, says the first out of the side of his mouth, and the others agree. Hours pass, then movement, then everything starts moving, they're all pulled away, some great power pulls them up, upward and upward in the direction of the sky, which soon breaks and opens onto another world full of peculiar fish. (Stefánsson 2010, 66)

Who these peculiar fish are Stefánsson does not say. They could be other fish species, or the humans that caught them. Or are they simply the cod themselves, made strange to each other when robbed of their aquatic element and removed to a world that will, eventually, be their death? Such estranged cod are, in any case, what this book is about. They are fish that roam an ocean increasingly crowded by new and expansive industries, that are drawn into domestication, into innovation, that are valued and ordered as market commodities, and that are put forward to the consideration of Parliament. We have seen the cod perform as biology, resource, science object, raw material, commodity, and capital, it has embodied policy

FIGURE 8.1
Illustration of the NOK 200 banknote, issued in 2017 by the Bank of Norway.

documents and procedures, and been dished up, steaming hot and on a dinner plate. What the cod that we write about have in common, however, is that the purpose of their estrangement is economic. Like the cod swimming across the example of a NOK 200 bill in figure 8.1, they are being drawn into various versions of economization. Ceaselessly and stubbornly, and not unlike the oceans living beings, the economies we study seek to grow and become larger, the sum of their efforts amounting to what we can today identify as the great economization of the ocean.

In the making and writing of this book we have traveled far out to sea, to distant and foreign markets, and into state bodies. Moving with and within a changing ocean landscape, we have shown how an ocean economy is made to emerge—carefully, deliberately, sometimes modestly, but other times quite strongly—whereby the ocean's wealth as well as its worth are presented in a myriad of ways: by way of maps that order and materialize the ocean's great economization; by domestication practices that institute cod bodies and stocks as biocapital; by insisting that nature is a site and a place of its own, with a production system working toward its own ends; by demands to take this nature into account, not only as on object of innovation, a problem, and a challenge to be solved but as a system of wealth and production; and by efforts to meet such demands through more-than-market valuation arrangements and tools of valuation that we have suggested thinking of as prices for collective concerns.

FROM FOSSILS TO FISH: A NEW PLATFORM ECONOMY?

To grow economies that take the living as their main input and resource, this book has shown, is not a simple task. Even for the salmon, whose domestication has progressed far more than the cod's, problems with diseases and animal welfare have put a stop to expansion; salmon farmers have in recent years turned huge profits but have also been prevented from growing at their desired pace. To overcome this, the traditionally low-tech salmon farming industry has worked to develop new production methods, a key aim being to establish production facilities at offshore locations. Several prototypes for such facilities have been produced, a main feature of these being their enormous size. For instance, the first of the so-called offshore platforms to have been put in operation, the "SalMar Ocean Farm 1 platform," holds up to 1.5 million salmon, while the legal maximum of a regular net pen is 200,000 (SalMar ASA 2019; IntraFish 2018). Moving farther out to sea, the industry is reaching for a production volume that is massively scaled up. Following the naming of these offshore production facilities as "platforms," one could even speculate if it aims to be the *new* platform industry, in company with, or perhaps a replacement of, the platforms of the petroleum industry.

In more modest ways and numbers, the promise of producing an increased output from the ocean is also present for the cod. For despite cod farming having failed repeatedly, the work to breed and improve the cod has continued, the belief being that one day, "the biology" will be "good enough." And indeed, investments in cod farming are once again beginning to pick up. For instance, the company Statt Torsk, which was established in 2014 by a group of minor investors, was, in 2021, priced at NOK 300 million, before raising its equity capital by bringing in NOK 115 million (Furuset 2021). As stated on the company's website, "Statt Torsk will forge a new frontier for Norwegian aquaculture by farming cod. We will lead the way in developing a new, sustainable industry based on selling high-quality products to markets in Norway and the world" (Statt Torsk, n.d.). By 2025, it aims to slaughter 12,000 tons of cod, which equals about 6 percent of the 2021 export of ocean-captured cod. Perhaps one day, the fishers will find that it is not petroleum platforms that displace them from their fishing grounds, but platform-reared cod, cod that will also crowd their markets.

Not only aquaculture is being scaled up and moved out to deep-sea waters, but also the fisheries. For as the tradable fishing quotas are becoming increasingly expensive they are to a greater degree held by large and capital-intensive oceangoing vessels, and to a lesser extent by the owners of smaller vessels fishing near the coast. This has provoked much political debate as well as a new social movement—Kystopprøret—"the coastal rebellion" that grew out of the city of Vardø, and so the very same place as where the struggle for minimum prices culminated in 1937. The current state of Vardø is also in and of itself testimony to how crucial the fisheries are to coastal communities. With the Barents Sea within immediate reach, it has all-year access to the great stocks of *skrei*, but in the absence of quotas and fish being landed on its shores it has experienced severe depopulation and is, despite the ardent efforts of remaining inhabitants, in a state of abandonment and dereliction. As much as it is about the right to fish, the coastal rebellion's struggle is about the right to live by, with, and of the ocean.

If we were to follow the cod beyond the cases explored in this book, it would likely take us into an examination of how the economies being struggled over come about. For it is with this, we find, that the contributions of the book lie—with opening the study of economy to broader concerns and a broader set of practices and agencies than those that models, calculations, structures, rules, mechanisms, predetermined logics, and premises can capture. It is our hope that through this we can entice other scholars perhaps not so versed in economics, or in the classical debates of economic sociology, but who nonetheless are confronted by their own versions of economization—be it in nature, in the city, in science, or somewhere else entirely—to pursue their own analyses of how these come about unfold, and the valuations they entail. This is a start, then, toward what we suggest thinking of as a form of empirical economy.

TOWARD THE STUDY OF EMPIRICAL ECONOMY

Throughout our work with this book, we have engaged closely with the concepts and approaches of social studies of markets and valuation studies. As we describe in chapter 1, these fields deserve a lot of credit for opening the economy to being studied as a practice and not a preexisting entity to be "discovered" and described by scientific method, preferably economics.

And yet, as we have sought to work with these literatures in our explorations of an economy whose very existence is premised on nature's affordances and capacities, we have also found that we needed to add to and extend their analytical lenses as well as vocabularies. Mainly three things have been addressed. First, that economization is not *one* specific type of process taking place at multiple locations, but rather as multiple *versions of economization* playing out at different times. Sometimes these versions overlap seamlessly, other times they clash and compete. Second, and much through our close engagement with the Atlantic cod and its many ways of resisting or altogether escaping economization, what we find, rather than a one-way progressive move toward something becoming more economic, are the *co-modifications* of nature and economy—the mutual modifications of entities of nature and practices of economy that take place when these are brought together and meet. Economization is a two-way street, at least. Third, we have found that various operations of the state are crucial parts of what economizations are about—the state's "long arms" sometimes being extended in the form of research councils and innovation agencies, other times by bringing the cod and its economization to the consideration of Parliament and ministries. Indeed, the state is found to be an active participant in product development and branding, and in the building of market architectures and designs. Consequently, we are confronted with a pressing hybridity—be it by entities that theory assumes are passive and which are instead active; by marketing work that is assumed to belong well within the bounds of private enterprise but is instead performed by government agencies; or by imperatives that supersede what is often perceived as the number-one goal of economization—to make a profit and generate surplus value—and are instead about valuing and maintaining a way of settlement and life along the coast. This means that we need concepts that help keep our analyses sufficiently close to empirical complexity; that can move across and hold together such not-so-purely economic economizations; and still be geared toward capturing the specificities of economic practice.

The notion of versions of economization has helped us to open up such an analytical space, simultaneously holding on to how these versions are ordered and patterned, and thereby express distinct economizations: of a seabed ordered into grids made available to the petroleum industry; of an exclusive economic zone and fishery commons cut through by historical rights and new quota regulations; of an experimental scientific constitution of the cod as biocapital, drawn into not one, but several value orderings of a future aquaculture industry; of an innovation paradigm that redefines failure as a learning opportunity, sees no limits to growth, and takes nature *into* innovation; and of prices and market architectures and designs that serve not only markets, but also collective concerns. To detect and identify how these economizations work and operate, the notion of *little tools of valuation* comes out as particularly helpful. The "little" reminds us to stay with the everyday, mundane work of economic practices, but does not prevent us from examining how these tools, when tied in with the valuation apparatuses of the political as well as of markets, become key to the valuations and orderings of economy. Furthermore, the term is useful because it opens analysis to the wider terrain of tools that we have found at work in the cases we examine. The tools of valuation that we describe may well simultaneously be tools of the political and tools of science, but their importance to and centrality in economizations is not lesser. Tied in with *valuation arrangements* and important also to the constitution of what we have suggested thinking of as *value orderings*, the tools of valuation that we examine are involved in valuing and ascribing worth, in assessing what is worth doing, in giving directions for what good order is, in praising some things and devaluing others, and in placing things in the preferred order. And so, by seeking to bring a nature economy into being—an *ocean* economy—they enact and engage with a far richer reality than that which enters the calculative spaces and models of economics. They engage with normativity and the political, with entities of nature and their intricate habitats, and with how the world is and should be. It is to this richness, or "mess," as John Law (2004) would have put it, that we need to attend if we are to understand an economization as great as that of the ocean.

Having followed the Atlantic cod as it has been drawn into and made part of the economization of the ocean, we have taken this species as our object to trace and think with, to learn from and reenact throughout the chapters of the book. In fact, as this book has shown, there is not *one* great economization of the ocean, as such economizations play out at multiple sites and come in quite different versions. There is the complex transformation from cod to capital and, more generally and overridingly, that of transforming nature into economy. And attached to all of that, there is the quite spectacular transformation that goes far beyond the economic: tiny cod eggs that (if all goes well) drift with the water and feed off creatures that came into life during the same spring flowering as they, that metamorphose into perfect little cod fish, grow and mature sexually, and venture across oceans only to reproduce and then return to their Barents Sea habitat.

A different animal or ocean materiality would likely take us in other directions, and to questions yet unexplored. Our study is moreover situated in the oceans, cultures, and politics of Norway, and so the relationship between economy and nature described in this book is, too. The worth of nature, as Marion Fourcade (2011) so strongly demonstrates, is not universal, but *situated*; it may shift over time and with the very tools and apparatuses we use to value it. Still, like we learn from the French and US cases explored by Fourcade, the contributions also of this book can carry across to other scholars interested in the relationship between nature and economy and, more generally, in conducting a form of *empirical economy*. Not so unlike the notion of empirical philosophy put forward by Annemarie Mol (2002), we take this to signify an analytics that is grounded not in predetermined principles about what economy is, but that seeks to understand economy by way of how it comes about in practice—and so also what it does. In its most basic sense, empirical economy is about studying economy as it unfolds and comes into being. It is about attending to the practices of economy, tracing these as versions of economization emerge, unfold, stabilize, are disrupted, even destroyed—and about how this takes the form of patterned realities and, consequently, has real-life consequences. This then also brings us to one last question, or issue to address, and which although it surfaces in many of the tools, arrangements, and orderings discussed throughout the book, is also quite elusive and only rarely spelled out: How is all this growth justified? How is it reasoned to be, in and of itself, good?

THE VALUE OF A LIFE ECONOMIC

The question of the good is a long-standing concern in the study of economy. For instance, in the early eighteenth-century publication *The Fable of the Bees; or, Private Vices, Publick Benefits*, Bernard Mandeville ([1924] 1988) argues that to generate wealth is not a moral virtue and good in and of itself. Instead, to Mandeville it is a vice, and so for an individual's pursuit of wealth to be legitimate, it must serve a greater, common good. With any aspiration to growth, justification must therefore be made. In the highly appraised work of Luc Boltanski and Eve Chiapello, *The New Spirit of*

Capitalism (2005a), the question of the common good is yet again raised, as they, with their twenty-first-century version, make an argument quite like Mandeville's. "It is impossible for capitalism," Boltanski and Chiapello (2005b, 163) write, "to avoid being at least somewhat oriented toward the attainment of the common good, as it is this striving which motivates people to become committed to its process." In both practice and thinking, economy has been involved in questions concerning the ethical and normative, of what is good and worthy—be it by way of a Protestant ethic (Weber [1904] 2001), a moral economy *outside* capitalist relations (Thompson 1971), a moral economy of its own *inside* economy (Fourcade 2017), or an economy put forward as in itself being and doing good: a "good economy" (Asdal et al. 2021).

Throughout the cases that this book explores, growth is seldom justified with reference to one, great common good. There is no "spirit" that asserts itself by way of reasoning or explicit discussion of what growth should serve. Indeed, perhaps the most direct address is made in the appeal for *modest* growth that we examined in chapter 4, by the parliamentarian asking for a cautious approach to growth, but simultaneously stating that "Industrial growth in coastal Norway is a notion that rings well in the rural areas" (Parliamentary Proceedings 1985–1986, 626–627). For "coastal Norway," this statement asserts, growth is of the good—a good that is also affirmed by the valuation arrangement of minimum-prices discussed in chapter 6, and how this, when it was implemented, sought to secure the fisheries as a viable way of living along the coast. The good operates in other, and less explicit ways, too. For instance, in one of the document species we have engaged quite closely with, the NOU report, we have seen how these work as tools of valuation that ascribe worth to things, be they a fishery commons or the nature set to carry aquaculture, thereby also considering what is worth doing and what the good manner of proceeding is. This suggests that what we have addressed in this book as value orderings also enact the good, not always explicitly or by stated reasoning, but by way of ordering actions, relations, and practices. Such implicit enactment of the good also comes to view in the qualitative constitution of capital explored in this book. As we show in chapter 5, investment capital is within the innovation paradigm asked to take on qualities considered as good to the cod farming industry—patience, competence, and risk-willingness. Cod constituted as biocapital, embedded a wholly different type of good, however, as chapter 3 shows.

Raised and reared as stocks, rather than being considered as fish of individual value, biocapitalization encapsulates the idea that capital should always be put to work and accumulate, and thereby the very idea that growth is of the good. We also observe how caring for the good is often delegated to "someone" or "somewhere" else: in the innovation paradigm the consumer is considered or made to carry the concern of the ethical and sustainable; in efforts to withstand growth at large scales and at a quick pace, nature is equipped for the task of standing up against this by imbuing it with distinct ecological qualities.

We could continue this list of examples, the point being that once we start to ask what the good implied by the valuations performed in economization processes is, we encounter not *the* common good but rather multiple versions of it, each of the versions of economization that we identify throughout the book also holding its own justification of what ends it serves. As Charis Thompson observes in her book *Good Science* (2013), the science she here follows is shot through with ethics. The same, we suggest, goes for economy. We have suggested the notion of "good economy" as an analytical tool to pursue economy's relations to versions of good (Asdal et al. 2021), and we also, for future studies, suggest the notion of value orderings to be further developed toward grasping this. There is quite clearly a strong relationship between how the valuation of entities and relations is ordered and how an economy purports to be and produce the good. Delineating not only how valuations are done, but also the compositions and registers of value is a task at hand.

Having followed the cod over more than seven decades and into several versions of economization, one of the things that strikes us is how much work—scientifically, politically, economically, and practically—is entailed to profit from it. Much of this work can be captured as seeking to instill in the cod the qualities of a nature that is good to the economy—to be a fish that conforms to cost-effective, efficient, and large-scale modes of producing, transporting, and marketing. Imbuing the cod with these desired qualities is an important part of constituting it as a good and thereby of its commodification, which in turn subjects it to certain rationalities or economic imperatives—to be governed, distributed, profited on, and have its value maximized. Still, and as we have firmly established in this book, the work to constitute the cod and its ocean environments as a nature that conforms to and is good to economy is not always met by a cod that is performing the role of "good nature." There is work that fails and is not of the good, as millions get lost, not only from the bank accounts of investors, but also in the form of cod who suffer and die when economization goes wrong.

By grounding our study in the empirical realities of the cod—be they living or dead, on paper or in the flesh—we have showed how the great economization of the ocean is accompanied by massive disorder as well. There is trouble, failure, disease, loss, and death. Surely, much of the work that we have described in this book is about something as simple, and difficult, as keeping things alive. Be it laboratory life or shelf life; the life of biocapital; life in the ocean; life along the coast; life in pools, ponds, and net pens; even market life—it requires work, care, knowledge, and—as the innovation paradigm also asks of its investment capital—quite a lot of patience.

This book has explored how the domesticated cod gets constituted as a form of biocapital, a capital whose accumulation is hinged on its reproduction and rearing. It is a creature whose life represents a future price to be achieved in markets, but whose being alive is a cost, and so being alive is justified by its productivity solely; it deserves its existence by being a source for economic surplus. Its life is genuinely economical, praised for the selling price it can achieve. However, it is also a "wild" entity, not only made to swim for the benefit of a "this-worldly" economy, but swimming also for its own sake. The cod of "the wild" is equally considered a stock, albeit a finite resource in a different sense: it is one that must be managed; there are rights attached to fishing it; and so perhaps more than the domesticated cod it is conceived as a good in itself, its existence a testimony to the condition of the cod stocks. It is part of an ocean commons, a sea with marvels, creatures, sounds, and secrets—appraised for its benefits and services, but also for the wonder of its shared existence. And so, it is perhaps fitting to end this book by asking how well our perceptions of the common good today capture species like the Atlantic cod, a being that must also thrive and live well, be cooperative and lend itself to us, if its economization is to continue.

NOTES

INTRODUCTION

1. We write "the wild" in quotation marks to signal that we do not take this to represent a category of purity or absence of the social. The cod is a species that has been fished for thousands of years and is today one that swims in environments subject to much human interference. As such, there are no cod that are truly wild, nor is the ocean a place of pure wilderness. Rather, and much in line with the work of Sarah Whatmore (2002, 12), we think of the cod and the ocean as a "promiscuous" topology of wildlife, one that is surreptitiously hybrid, and that works to "render the experience of radical difference delineating the human from the animal, the civilized from the wild, as a con-figuring—a drawing together, as Jennifer Ham puts it—rather than a holding apart."

CHAPTER 2

1. For further discussion on the normative aspects of emerging economies, see Asdal et al. 2021.

2. "Det system som foreslås for utlysning av områder ligger nær opp til det som er valgt for Storbritannia (jfr. Statutory Instruments 1964 No 708.—Continental Shelf Petroleum—Schedule 2). Systemet går igjen i en rekke moderne oljelover" (p. 30, Innst. 1965: 44. "Innstilling om utforskning og utnyttelse av undersjøiske petroleumsforekomster med utkast til kongelig resolusjon, Innstilling nr. 2 fra Utvalget til å foreslå regler om utforming av undersjøiske naturforekomster").

3. "Resolution on Norwegian Sovereignty over Certain Underwater Areas," Ministry of Petroleum and Energy, accessed March 21, 2022, https://lovdata.no/dokument/SF/forskrift/1963-05-31-1.

4. "50 Years since the First Licensing Round," Oljedirektoratet, last modified November 27, 2015, https://www.npd.no/fakta/nyheter/generelle-nyheter/2015/50-ar-siden-forste-konsesjonsrunde/.

5. "50 Years since the First Licensing Round."

6. Translation by the authors, original inscription: "KART SOM VISER TILDELING FOR UTVINNINGSTILLATELSE FOR PETROLEUM."

7. See Marx, *Capital* ([1867] 2018), chapter 27.

8. Today, there are two different licensing systems in use: "On the Norwegian continental shelf, there are two equal types of licensing rounds that will ensure an efficient and rational exploration of the entire Norwegian continental shelf. Those are awards in predefined areas (APA) for mature parts where the knowledge is highest and numbered licensing rounds in other areas. This ensures that all parts of the Norwegian continental shelf can be adequately explored. All areas that are open and accessible for petroleum activity may be announced in a licensing round." "Exploration Policy," Norwegian Petroleum, last modified March 22, 2022, https://www.norskpetroleum.no/en/exploration/exploration-policy/.

9. As stated by the *United Nations Convention on the Law of the Sea*, Article 5: "Except where otherwise provided in this Convention, the normal baseline for measuring the breadth of the territorial sea is the low-water line along the coast as marked on large-scale charts officially recognized by the coastal State" (UN 1982, 27).

10. As stated by the *United Nations Convention on the Law of the Sea*, Article 5: "Except where otherwise provided in this Convention, the normal baseline for measuring the breadth of the territorial sea is the low-water line along the coast as marked on large-scale charts officially recognized by the coastal State" (UN 1982, 27).

11. This was doubled after the later declaration of Norwegian fishery zones around Svalbard (1976) and Jan Mayen (1980). In total, the three zones now cover more than two million square kilometers, which is more than six times the Norwegian mainland territory (Christensen 2014a, 54); see also Edgeir Benum, "Da havretten femdoblet Norge [When the law of the sea quintupled Norway]," Norgeshistorie (Norwegian History), accessed July 30, 2021, https://www.norgeshistorie.no/oljealder-og-overflod/1938-da-havretten-femdoblet-norge.html.

12. "Act on the Management of Wild Marine Resources (Marine Resources Act)," Ministry of Trade and Industry, accessed February 4, 2022, https://lovdata.no/dokument/NL/lov/2008-06-06-37#KAPITTEL_1; Marine Resources Act, chapter 1, paragraph 2: "Dei viltlevande marine ressursane ligg til fellesskapet i Noreg" (translated by the authors).

13. "Act on the Management of Wild Marine Resources"; Marine Resources Act.

14. "FELT," Norwegian Petroleum, accessed July 30, 2021, https://www.norskpetroleum.no/fakta/felt/.

15. "White Papers," Norwegian Government Security and Service Organisation, accessed February 14, 2022, https://www.regjeringen.no/en/find-document/white-papers-/id1754/.

16. This equals around NOK 135 million in 2021, or around 13 million Euro.

CHAPTER 3

1. Curiously, what initially caught our interest in these experiments toward cod domestication was that much of the scientists' work was funded by the Oil–Fish Fund. In fact, it has been suggested that the motivation for the project was not primarily aquaculture, but to create a kind of "backup cod" (Aarset 2005), a cod that in the case of a catastrophic oil spill could be bred in large numbers and subsequently put out into the ocean, thus replacing the stocks wiped out by the oil. Still, having delved into the archives of the Oil–Fish Fund, read the scientists' reports and research papers, and interviewed several of the scientists involved, we have found nothing that indicates that the idea of such a backup cod existed at the time.

2. Interview with former marine biologist at Flødevigen Biological Station, September 24, 2019; and with former marine biologist at the Flødevigen Biological Station, Austevoll Marine Aquaculture station, and fish farming entrepreneur, October 22, 2019.

3. Interview with former marine biologist at the Flødevigen Biological Station, Austevoll Marine Aquaculture station, and fish farming entrepreneur, October 22, 2019.

4. Interview with former marine biologist.

5. Interview with former marine biologist.

6. Interview with former marine biologist.

7. Interview with former marine biologist.

8. Interview with marine biologist at the Austevoll Marine Aquaculture Station, September 23, 2019; and with former marine biologist at the Flødevigen Biological Station, Austevoll Marine Aquaculture station, and fish farming entrepreneur, October 22, 2019.

9. Interview with former marine biologist at the Flødevigen Biological Station, September 24, 2019.

CHAPTER 5

1. The Norwegian Seafood Council was at the time named the Export Agency for Seafood. It was given its current name in 2012.

2. Interview with former investor in and manager of cod farming company, May 31, 2017; and with fish farm manager and former investor and owner in cod farming, June 15, 2017.

3. Interview with cod farmer and investor, May 31, 2017.

4. Interview with former investor in and manager of cod farming company, May 31, 2017.

5. Interview with cod farmer and investor, June 15, 2017.

CHAPTER 6

1. For an overview of these literatures and their engagement with price, see Çalışkan (2009, 2010).

2. Interview and observation at Myre fish landing station, March 7, 2017.

3. Interview and observation at Myre fish landing station.

4. The Norwegian Fishermen's Sales Organization has over the years evolved into becoming a rather complex organization that describes itself as conducting the simultaneous tasks of "representation, business, and public administration" (interview with representative of the Norwegian Fishermen's Sales Organization, February 8, 2017): it works to advance the interests of the fishers; it functions as an intermediary for payment between the firsthand buyers and the fishers; it collects data on the trade between fishers and the firsthand buyers; and it carries out duties on behalf of the government. These duties include routine controls at the fish landing stations to ensure that they are not underreporting how much fish is coming in and to register the quality of the fish being landed. Also, and as we shall return to later in this chapter, the Norwegian Fishermen's Sales Organization collaborates with another state-mandated organization, the Norwegian Seafood Council, to control the use of the Skrei quality brand. Its most important task, however, is to administer and oversee the minimum price system.

5. "Act on Firsthand Sale of Wild Marine Resources (Fiskesalslaglova)," Ministry of Trade and Industry, accessed June 30, 2021, https://lovdata.no/dokument/NL/lov/2013-06-21-75?q=r%C3%A5fiskloven.

6. "Act on Firsthand Sale of Wild Marine Resources."

7. "Act on the Management of Wild Marine Resources (Marine Resources Act)," Ministry of Trade and Industry, accessed June 30, 2021, https://lovdata.no/dokument/NL/lov/2008-06-06-37?q=havressursloven.

8. Interview with fish trader, February 7, 2017.

9. Interview with daily manager of fish landing station, March 7, 2017.

10. Interview with owner and captain of coastal fishing vessel, March 23, 2017.

11. Interview with owner and captain.

12. Interview with daily manager of fish landing station, March 7, 2017.

13. Interview with daily manager.

14. "Regulations on the Quality of Fish and Fish Products," Ministry of Trade and Industry, accessed March 26, 2022, https://lovdata.no/dokument/SF/forskrift/2013-06-28-844.

15. Interview with head of Skrei brand, Norwegian Seafood Council, January 27, 2017.

16. Interview with head of Skrei brand.

17. Interview with head of Skrei brand.

18. Interview with head of Skrei brand.

19. Interview with member of the Skrei Patrol, Norwegian Fishermen's Sales Organization, April 5, 2017.

20. Interview with member of the Skrei Patrol.

21. Interview with member of the Skrei Patrol.

CHAPTER 7

1. The list of ministers visiting China was provided by the Ministry of Foreign Affairs (May 12, 2020); the list of businesses, institutions, and organizations accompanying the Minister of Fisheries and the Minister of Trade and Industry was provided by the Ministry of Trade, Industry, and Fisheries (May 20, 2020).

2. "Act on Regulation of Exports of Fish and Fishery Products [Fish Export Act]," Ministry of Trade and Industry, accessed June 26, 2021, https://lovdata.no/dokument/NL/lov/1990-04-27-9.

3. Minutes from parliamentary debate, Ot.prp. No. 90, 1988–1999, March 29, 1990, https://stortinget.no/no/Saker-og-publikasjoner/Saker/Sak/?p=6640.

4. Today this is the Ministry of Trade, Industry and Fisheries.

5. Minutes from parliamentary debate, Ot.prp. No. 90, 1988–1999; Ot.prp. No. 88, 2004–2005; Ot.prp. No. 90, 1988–1989: 13, March 29, 1990, https://lovdata.no/dokument/NL/lov/1990-04-27-9.

6. Ot.prp. No. 9, 1988–1989: 9; §6. Oppgaver for Sjømatrådet, Fiskeridirektoratet og tollmyndighetene, in Forskrift om regulering av eksporten av fisk og fiskevarer. Available at https://lovdata.no/dokument/SF/forskrift/1991-03-22-157.

7. Interview with head of white fish market group, Norwegian Seafood Council, November 14, 2016.

8. Interview with financial advisor, Norwegian Seafood Council, August 25, 2017.

9. "Market Groups," Norwegian Seafood Council, as of June 9, 2020; https://seafood.no/om-norges-sjomatrad/markedsgrupper/.

10. "Offices and Priority Markets," Norwegian Seafood Council, as of June 9, 2020; https://seafood.no/om-norges-sjomatrad/kontorer/.

11. Foreign Service Act, last amendment in 2017—§4. Other personnel posted abroad: "Other personnel posted abroad are officials who through a decision by the Ministry are attached to a mission for a fixed period of time, and who are employed by independent legal entities. Other personnel posted abroad are answerable to the head of mission." https://lovdata.no/dokument/NLE/lov/2015-02-13-9 (accessed June 27, 2021).

12. "Team Norway," Ministry of Trade and Industry, accessed June 26, 2021, https://www.regjeringen.no/no/tema/naringsliv/internasjonalt-naringssamarbeid-og-eksport/team-norway/id2344658/.

13. "Prop. 11 L (2014–15), ch. 5.8.1," Ministry of Foreign Affairs, accessed June 26, 2021, https://www.regjeringen.no/no/dokumenter/Prop-11-L-20142015/id2009323/sec5.

14. Interview with head of China office, Norwegian Seafood Council, March 8, 2018.

15. Interview with head of Skrei Quality Brand marketing, Norwegian Seafood Council, July 6, 2017.

16. Interview with head of China office, Norwegian Seafood Council, March 8, 2018.

17. Interview with head of China office.

18. Interview with freezer hotel manager, February 2, 2017; and interview with trawler manager, February 9, 2017.

19. Interview with head of China office, Norwegian Seafood Council, March 8, 2018.

20. Interview with head of China office.

21. Interview with head of China office.

22. Interview with head of China office.

23. Interview with head of China office.

24. Interview with head of China office.

25. Interview with head of China office.

REFERENCES

Aarset, Bernt. 2005. "Sjødyr på beite eller eksklusiv fangst? Regulering og modeller for havbeite i den norske kystsonen. Forprosjekt om havbeite. Delprosjekt II: kartlegging av problemstillinger knyttet til to havbeiteprosjekter [Marine animal grazing or exclusive catch? Regulation and models for marine grazing in the Norwegian coastal zone. Preliminary work on marine grazing. Subproject II: Mapping of problems regarding marine grazing projects]." *SNF-report nr. 07/05*. Bergen: Samfunns- og næringslivsforskning.

Adler, Antony. 2019. *Neptune's Laboratory: Fantasy, Fear, and Science at Sea*. Cambridge, MA: Harvard University Press.

Alexandersen, Rune. 2014. "Marine Harvest Will Withdraw 70 Million from the Seafood Council and Spend It on Its Own Marketing." *Nordlys*, November 18, 2014. https://www.nordlys.no/marine-harvest-vil-trekke-70-millioner-ut-av-sjomatradet-og-bruke-pa-egen-markedsforing/s/5-32-6019.

Anderson, David G., Jan P. L. Loovers, Sara A. Schroer, and Robert P. Wishart. 2017. "Architectures of Domestication: On Emplacing Human–Animal Relations in the North." *Journal of the Royal Anthropological Institute* 23 (2): 398–416.

Antal, Ariane B., Michael Hutter, and David Stark, eds. 2015. *Moments of Valuation: Exploring Sites of Dissonance*. Oxford: Oxford University Press.

Arbo, Peter, Maaike Knol, Sebastian Linke, and Kevin St. Martin. 2018. "The Transformation of the Oceans and the Future of Marine Social Science." *Maritime Studies* 17 (3): 295–304. https://doi.org/10.1007/s40152-018-0117-5.

Asdal, Kristin. 1998. *Knappe ressurser. Økonomenes grep om miljøfeltet*. Oslo: Universitetsforlaget.

Asdal, Kristin. 2003. "The Problematic Nature of Nature: The Post-constructivist Challenge to Environmental History." *History and Theory* 42 (4): 60–74.

Asdal, Kristin. 2008. "On Politics and the Little Tools of Democracy: A Down-to-Earth Approach." *Distinktion: Journal of Social Theory* 9 (1): 11–26.

Asdal, Kristin. 2011. "The Office: The Weakness of Numbers and the Production of Non-Authority." *Accounting, Organizations and Society* 36 (1): 1–9.

Asdal, Kristin. 2012. "Contexts in Action—And the Future of the Past in STS." *Science, Technology, & Human Values* 37 (4): 379–403. https://doi.org/10.1177/0162243912438271.

Asdal, Kristin. 2014. "From Climate Issue to Oil Issue: Offices of Public Administration, Versions of Economics and the Ordinary Technologies of Politics." *Environment and Planning A* 46 (9): 2110–2124. https://doi.org/10.1068/a140048p.

Asdal, Kristin. 2015a. "What Is the Issue? The Transformative Capacity of Documents." *Distinktion: Journal of Social Theory* 16 (1): 74–90. https://doi.org/10.1080/1600910X.2015.1022194.

Asdal, Kristin. 2015b. "Enacting Values from the Sea: On Innovation Devices, Value Practices and the Co-Modification of Markets and Bodies in Aquaculture." In *Value Practices in the Life Sciences and Medicine*, ed. Isabelle Dussauge, Claes-Fredrik Helgesson, and Francis Lee, 168–185. Oxford: Oxford University Press.

Asdal, Kristin. 2019. "Is ANT Equally Good in Dealing with Local, National and Global Natures?" In *The Routledge Companion to Actor-Network Theory*, ed. Anders Blok, Ignacio Farías, and Celia Roberts, 337–344. London: Routledge.

Asdal, Kristin, Brita Brenna, and Ingunn Moser. 2007. *The Politics of Interventions*. Oslo: Unipub.

Asdal, Kristin, and Béatrice Cointe. 2021. "Experiments in Co-Modification: A Relational Take on the Becoming of Commodities and the Making of Market Value." *Journal of Cultural Economy* 14 (3): 280–292.

Asdal, Kristin, Béatrice Cointe, Bård Hobæk, Hilde Reinertsen, Tone Huse, Silje R. Morsman, and Tommas Måløy. 2021. "'The Good Economy': A Conceptual and Empirical Move for Investigating How Economies and Versions of the Good Are Entangled." *BioSocieties*. https://doi.org/10.1057/s41292-021-00245-5.

Asdal, Kristin, Liliana Doganova, and Maximilian Fochler. Forthcoming. "Valuation Studies." In *STS Encyclopedia*, ed. Ulrike Felt and Alan Irwin.

Asdal, Kristin, and Tone Druglitrø. 2016. "Modifying the Biopolitical Collective." In *Humans, Animals and Biopolitics: The More-Than-Human Condition*, ed. Kristin Asdal, Tone Druglitrø, and Steve Hinchliffe. London: Routledge.

Asdal, Kristin, Tone Druglitrø, and Steve Hinchliffe. 2017. *Humans, Animals and Biopolitics: The More-Than-Human Condition*. Abingdon: Routledge.

Asdal, Kristin, and Bård Hobæk. 2016. "Assembling the Whale: Parliaments in the Politics of Nature." *Science as Culture* 25 (1): 96–116.

Asdal, Kristin, and Helge Jordheim. 2018. "Texts on the Move: Textuality and Historicity Revisited." *History and Theory* 57 (1): 56–74.

Asdal, Kristin, and Hilde Reinertsen. 2022. *Doing Document Analysis: A Practice-Oriented Method*. London: Sage Publications.

Aspers, Patrik, and Jens Beckert. 2011. "Value in Markets." In *The Worth of Goods: Valuation and Pricing in the Economy*, ed. Jens Beckert and Patrik Aspers, 3–40. Oxford: Oxford University Press.

Barbesgaard, Mads. 2018. "Blue Growth: Savior or Ocean Grabbing?" *Journal of Peasant Studies* (1): 130–149.

BarentsWatch. 2018. "Norway's Maritime Borders." Last modified October 9, 2018. https://www.barentswatch.no/en/articles/Norways-maritime-borders/.

Barry, Andrew, and Don Slater. 2002. "Introduction: The Technological Economy." *Economy and Society* 31 (2): 175–193.

Barua, Maan. 2016. "Lively Commodities and Encounter Value." *Environment and Planning D: Society and Space* 34 (4): 725–744.

Beckert, Jens, and Patrik Aspers, eds. 2011. *The Worth of Goods: Valuation and Pricing in the Economy*. Oxford: Oxford University Press.

Bendiksen, Bjørn Inge. 2018. "Referansepriser i førstehåndsmarkedet for hvitfisk. Faglig sluttrapport [Reference prices in the firsthand market for white fish. Scientific summary report]." Nofima report series.

Bennett, James Nathan, Jessica Blythe, Carole Sandrine White, and Cecilia Campero. 2021. "Blue Growth and Blue Justice: Ten Risks and Solutions for the Ocean Economy." *Marine Policy* (125). https://doi-org.ezproxy.uio.no/10.1016/j.marpol.2020.104387.

Besky, Sarah, and Alex Blanchette. 2019. "Introduction: The Fragility of Work." In *How Nature Works: Rethinking Labor on a Troubled Planet*, ed. Sarah Besky and Alex Blanchette. Albuquerque: University of New Mexico Press.

Bestor, Theodore C. 2004. *Tsukiji: The Fish Market at the Center of the World*. Berkeley: University of California Press.

Bingham, Nick, and Steve Hinchliffe. 2008. "Reconstituting Natures: Articulating Other Modes of Living Together." *Geoforum* 39 (1): 83–87.

Birch, Kean. 2017. "Rethinking Value in the Bio-Economy: Finance, Assetization, and the Management of Value." *Science, Technology, & Human Values* 42 (3): 460–490.

Bjerke, Espen. 2009. "Norway's Oil Production Is Plummeting." *Dagens Næringsliv*, December 15, 2009. https://www.dn.no/norges-oljeproduksjon-stuper/1-1-1413062.

Bjørkdahl, Kristian, and Tone Druglitrø, eds. 2016. *Animal Housing and Human-Animal Relations: Politics, Practices and Infrastructures*. London: Routledge.

Blaxter, John H. S. 1976. "Reared and Wild Fish: How Do They Compare?" In *Proceedings of the 10th European Symposium on Marine Biology, Ostend, Belgium, Sept. 17–23, 1975: 1*, ed. Guido Persoone and Edmonde Jaspers. Research in mariculture at laboratory- and pilot scale, 11–26. Wetteren: University Press.

Boldyrev, Ivan, and Ekaterina Svetlova, eds. 2016. *Enacting Dismal Science: New Perspectives on the Performativity of Economics*. New York: Palgrave Macmillan US. https://doi.org/10.1057/978-1-137-48876-3.

Boltanski, Luc, and Eve Chiapello. 2005a. *The New Spirit of Capitalism*. Translated by Gregory Elliott. London: Verso.

Boltanski, Luc, and Eve Chiapello. 2005b. "The New Spirit of Capitalism." *International Journal of Politics, Culture, and Society* 18 (3–4): 161–188.

Brattland, Camilla. 2013. "Proving Fishers Right. Effects of the Integration of Experience-Based Knowledge in Ecosystem-Based Management." *Acta Borealia* 30 (1): 39–59.

Braverman, Irus, and Elizabeth R. Johnson, eds. 2020. *Blue Legalities: The Life and Laws of the Sea*. Durham, NC: Duke University Press.

Brent, Zoe W., Mads Barbesgaard, and Carsten Pedersen. 2020. "The Blue Fix: What's Driving Blue Growth?" *Sustanability Science* (15): 31–43. https://doi.org/10.1007/s11625-019-00777-7.

Cahill, Damien, Melinda Cooper, Martijn Konings, and David Primrose, eds. 2018. *The SAGE Handbook of Neoliberalism*. Los Angeles: Sage.

Çalışkan, Koray. 2010. *Market Threads: How Cotton Farmers and Traders Create a Global Commodity*. Princeton, NJ: Princeton University Press.

Çalışkan, Koray, and Michel Callon. 2009. "Economization, Part 1: Shifting Attention from the Economy towards Processes of Economization." *Economy and Society* 38 (3): 369–398.

Çalışkan, Koray, and Michel Callon. 2010. "Economization, Part 2: A Research Programme for the Study of Markets." *Economy and Society* 39 (1): 1–32.

Callon, Michel. 1984. "Some Elements of a Sociology of Translation: Domestication of the Scallops and the Fishermen of St Brieuc Bay." *The Sociological Review* 32 (1 suppl.): 196–233.

Callon, Michel, ed. 1998. *The Laws of the Markets*. Oxford: Blackwell/The Sociological Review.

Callon, Michel. 2005. "Let's Put an End on Uncertainties." In "Quality: A Debate," ed. Christine Musselin and Catherine Paradeise, Supplement, *Sociologie du Travail* 47: e89–e123.

Callon, Michel. 2007. "What Does It Mean to Say That Economics Is Performative?" In *Do Economists Make Markets? On the Performativity of Economics*, ed. Donald MacKenzie, Fabian Muniesa, and Lucia Siu, 311–357. Princeton, NJ: Princeton University Press.

Callon, Michel, and John Law. 1982. "On Interests and Their Transformation: Enrolment and Counter-Enrolment." *Social Studies of Science* 12 (4): 615–625.

Callon, Michel, Cécile Méadel, and Vololona Rabeharisoa. 2002. "The Economy of Qualities." *Economy and Society* 31 (2): 194–217.

Callon, Michel, and Fabian Muniesa. 2005. "Peripheral Vision: Economic Markets as Calculative Collective Devices." *Organization Studies* 26 (8): 1229–1250.

Callon, Michel, Yuval Millo, and Fabian Muniesa, eds. 2007. *Market Devices*. Malden, MA: Blackwell.

Cassen, Christophe, and Béatrice Cointe. 2022. "From the Limits to Growth to Greenhouse Gas Emission Pathways: Technological Change in Global Computer Models (1927–2007)." *Contemporary European History*.

Cassidy, Rebecca. 2007. "Introduction: Domestication Reconsidered." In *Where the Wild Things Are Now: Domestication Reconsidered*, ed. Rebecca Cassidy and Molly Mullin, 1–25. London: Routledge.

Cassidy, Rebecca, and Molly Mullin, eds. 2007. *Where the Wild Things Are Now: Domestication Reconsidered*. London: Routledge.

Chiapello, Eve. 2007. "Accounting and the Birth of the Notion of Capitalism." *Critical Perspectives on Accounting* 18 (3): 263–296.

Chiapello, Eve. 2015. "Financialisation of Valuation." *Human Studies* 38 (1): 13–35.

Chiapello, Eve. 2020. "Financialization as a Socio-Technical Process." In *The Routledge International Handbook of Financialization*, ed. Philip Mader, Daniel Mertens, and Natascha van der Zwan, 81–91. London: Routledge.

Clutton-Brook, Juliet. 1989. *The Walking Larder: Patterns of Domestication, Pastoralism, and Predation*. London: Routledge.

Christensen, Pål. 2014a. "EF-strid og 200 mils økonomisk sone [EEC dispute and 200 mile economic zone]." In *Norges fiskeri- og kysthistorie bind IV: Havet, fisken og oljen 1970–2014* [Norway's fishery and coastal history vol. IV: The sea, the fish, and the oil 1970–2014], ed. Pål Christensen. Oslo: Fagbokforlaget.

Christensen, Pål. 2014b. "Fra overflod til overkapasitet [From abundance to overcapacity]." In *Norges fiskeri- og kysthistorie bind IV: Havet, fisken og oljen 1970–2014* [Norway's fishery and coastal history vol. 4: The sea, the fish, and the oil 1970–2014], ed. Pål Christensen. Oslo: Fagbokforlaget.

Clark, Nigel. 2003. "Feral Ecologies: Performing Life on the Colonial Periphery." *Sociological Review* 51 (2 suppl.): 163–182.

Cochoy, Franck. 2007. "A Sociology of Market-Things: On Tending the Garden of Choices in Mass Retailing." In *Market Devices*, ed. Michel Callon, Yuval Millo, and Fabian Muniesa, 109–129. Malden, MA: Blackwell.

Cole, H. S. D., Christopher Freeman, Marie Jahoda, and K. L. R. Pavitt, eds. 1973. *Thinking about the Future: A Critique of the Limits to Growth*. Sussex: Chatto & Windus for Sussex University Press.

Cooper, Melinda. 2008. *Life as Surplus: Biotechnology and Capitalism in the Neoliberal Era*. Seattle: University of Washington Press.

Cooper, Melinda, and Catherine Waldby. 2014. *Clinical Labor: Tissue Donors and Research Subjects in the Global Bioeconomy*. Durham, NC: Duke University Press.

Corner, James. 1999. "The Agency of Mapping: Speculation, Critique and Invention." In *Mappings*, ed. Denis Cosgrove, 213–252. London: Reaktion Books.

Dave, Naisargi N. 2019. "Kamadhenu's Last Stand: On Animal Refusal to Work." In *How Nature Works: Rethinking Labor on a Troubled Planet*. Albuquerque: University of New Mexico Press.

DeLoughrey, Elizabeth. 2016. "The Oceanic Turn: Submarine Futures of the Anthropocene." In *Humanities for the Environment: Integrating Knowledge, Forging New Constellations of Practice*, ed. Joni Adamson and Michael Davis, 242–258. London: Routledge.

DeLoughrey, Elizabeth. 2019. *Allegories of the Anthropocene*. Durham, NC: Duke University Press.

Demuth, Bathsheba. 2019. *Floating Coast: An Environmental History of the Bering Strait*. New York: W. W. Norton & Company.

Despret, Vinciane. 2016. *What Would Animals Say If We Asked the Right Questions?* Translated by Brett Buchanan. Minneapolis: University of Minnesota Press.

Dewey, John. 1927. *The Public and Its Problems*. London: George Allen & Unwin.

Dewey, John. 1939. *Theory of Valuation*. Chicago: University of Chicago Press.

Dobeson, Alexander. 2016. "Hooked on Markets: Revaluing Coastal Fisheries in Liberal Rural Capitalism." PhD diss., Copenhagen Business School.

Doganova, Liliana. 2018. "Discounting the Future: A Political Technology." *Economic Sociology. Perspectives and Conversations* 19 (2): 4–9.

Doganova, Liliana. 2019. "What is the Value of ANT Research into Economic Valuation Devices?" In *The Routledge Companion to Actor-Network Theory*, ed. Anders Blok, Ignacio Farías, and Celia Roberts, 256–263. London: Routledge.

Doganova, Liliana, and Marie Eyquem-Renault. 2009. "What Do Business Models Do? Innovation Devices in Technology Entrepreneurship." *Research Policy* 38 (10): 1559–1570.

Doganova, Liliana, and Peter Karnøe. 2015. "Clean and Profitable." In *Moments of Valuation: Exploring Sites of Dissonance*, ed. Ariane B. Antal, Michael Hutter, and David Stark, 229–248. Oxford: Oxford University Press.

Doganova, Liliana, and Fabian Muniesa. 2015. "Capitalization Devices: Business Models and the Renewal of Markets." In *Making Things Valuable*, ed. Martin Kornberger, Lise Justesen, Jan Mouritsen, and Anders K. Madsen, 109–127. Oxford: Oxford University Press.

Doganova, Liliana, and Vololona Rabeharisoa. 2022. "Price as an Epistemic and a Political Object: An Inquiry into 'the Most Expensive Drug Ever.'" In *Sub-Theme 64: Valuation and Critique in the "Good Economy."* Vienna: EGOS: 1–9.

Dooren, Thomas van, Eben Kirksey, and Urula Münster. 2016. "Multispecies Studies: Cultivating Arts of Attentiveness." *Environmental Humanities* 8 (1): 1–23.

Døssland, Atle. 2014. "DII 1: Kjøpmenn går inn i nye roller 1720–1815 [Part 1: Merchants enter new roles 1720–1815]." In *Norges fiskeri- og kysthistolie bind II: Eskpansjon i eksportfiskeria 1720–1880* [Norway's fishery and coastal history vol. II: Expansion in the export fisheries 1720–1880], ed. Atle Døssland, 15–280. Oslo: Fagbokforlaget.

Druglitrø, Tone. 2018. "'Skilled Care' and the Making of Good Science." *Science, Technology and Human Values*. Special Issue: *Science, Culture and Care in Laboratory Animal Research* 43 (4): 649–670.

Dussauge, Isabelle, Claes-Fredrik Helgesson, and Francis Lee, eds. 2015. *Value Practices in the Life Sciences and Medicine*. Oxford: Oxford University Press.

Eikeset, Anne Maria, Anna B. Mazzarella, Brynhildur Davíðsdóttir, Dane H. Klinger, Simon A. Levin, Elena Rovenskaya, and Nils Chr. Stenseth. 2018. "What Is Blue Growth? The Semantics of 'Sustainable Development' of Marine Environments." *Marine Policy* (87): 177–179.

Elliot, John E. 2008. "Introduction to the Transaction edition." In Joseph A. Schumpeter, *The Theory of Economic Development*. New Brunswick, NJ: Transaction Publishers.

Endresen, Rune. 2016. "The Bankruptcy Was the Start of a New Adventure: We Were Accused of Cheating from Day One." *Nord24*, November 15, 2016. https://www.nord24.no/vekst-i-nord/fiskeri/jobb/konkursen-ble-starten-pa-et-nytt-eventyr-vi-ble-beskyldt-for-juks-fra-dag-n/s/5-32-71857?key=2020-06-17T09:31:25.000Z/retriever/2a07529e2cf668912d27cf3a0b63bcf9e961125c.

Enoksen, Ken H. 2017. "Å temme torsken–fremveksten av norsk torskeoppdrettsnæring. En beretning om kollektivt entreprenørskap." PhD diss., UiT, Arctic University of Norway.

FishBase. n.d. "Yolk-Sac Larva." Accessed March 4, 2022. https://www.fishbase.se/glossary/Glossary.php?q=yolk-sac+larva.

Fisher, Irving. 1896. "What Is Capital?" *The Economic Journal* 6 (24): 509–534.

Fiskets Gang nr. 15/16 (32). 1985. Published by the Director of Fisheries.

Fligstein, Neil. 1996. "Markets as Politics: A Political-Cultural Approach to Market Institutions." *American Sociological Review* 61 (4): 656–673.

Fligstein, Neil. 2001. *The Architecture of Markets: An Economic Sociology of Twenty-First Century Capitalist Societies*. Princeton, NJ: Princeton University Press.

Forbrukerrådet. n.d. "Fairtrade." Accessed March 25, 2022. https://www.forbrukerradet.no/merkeoversikten/etikk/fairtrade/.

Fortun, Mike. 2008. *Promising Genomics: Iceland and deCODE Genetics in a World of Speculation*. Oakland: University of California Press.

Foucault, Michel. (1970) 1974. *The Order of Things: An Archaeology of the Human Sciences*. London: Routledge.

Foucault, Michel. (1978) 2007. *Security, Territory, Population: Lectures at the Collège de France, 1977–78*. Edited by Michel Senellart, Francois Ewald, and Alessandro Fontana. Translated by Graham Burchell. London: Palgrave Macmillan. https://doi.org/10.1057/9780230245075.

Fourcade, Marion. 2009. *Economists and Societies. Discipline and Profession in the United States, Britain, and France, 1890s to 1990s*. Princeton, NJ: Princeton University Press.

Fourcade, Marion. 2011. "Cents and Sensibility: Economic Valuation and the Nature of 'Nature.'" *American Journal of Sociology* 116 (6): 1721–1777.

Fourcade, Marion. 2017. "The Fly and the Cookie: Alignment and Unhingement in the 21st-Century Capitalism." *Socio-Economic Review* 15 (3): 661–678.

Frankel, Christian. 2018. "The 's' in Markets: Mundane Market Concepts and How to Know a (Strawberry) Market." *Journal of Cultural Economy* 11 (5): 458–475.

Frankel, Christian, José Ossandón, and Trine Pallesen. 2019. "The Organization of Markets for Collective Concerns and Their Failures." *Economy and Society* 48 (2): 153–174.

Franklin, Sarah. 2003. "Ethical Biocapital: New Strategies of Cell Culture." In *Remaking Life and Death: Toward an Anthropology of the Biosciences*, ed. Sarah Franklin and Margaret Lock, 97–127. Santa Fe, NM: School of American Research Press.

Franklin, Sarah. 2007. *Dolly Mixtures: The Remaking of Genealogy*. Durham, NC: Duke University Press.

Franklin, Sarah, and Margaret M. Lock, eds. 2003. *Remaking Life and Death: Toward an Anthropology of the Biosciences*. Santa Fe, NM: School of American Research Press.

Franklin, Sarah, and Helena Ragoné, eds. 1998. *Reproducing Reproduction: Kinship, Power, and Technological Innovation*. Philadelphia: University of Pennsylvania Press.

Freeman, Cristopher. 1973. "Malthus with a Computer." In *Thinking about the Future: A Critique of the Limits to Growth*, ed. H. S. D. Cole, Christopher Freeman, Marie Jahoda, and K. L. R. Pavitt, 5–13. Sussex: Chatto & Windus for Sussex University Press.

Friese, Carrie. 2015. "Genetic Value: The Moral Economies of Cloning in the Zoo." In *Value Practices in the Life Sciences and Medicine*, ed. Isabelle Dussauge, Claes-Fredrik Helgesson, and Francis Lee, 153–167. Oxford: Oxford University Press.

Furuset, Anders. 2021. "Hjørnestensinvestor solge nesten hIlvparen av aksjene sine i torskeoppdretter [Cornerstone investors sell almost half of their shares in cod farms]." IntraFish, May 31, 2021. https://www.intrafish.no/okonomi/hjornestensinvestor-solgte-nesten-halvparten-av-aksjene-sine-i-torskeoppdretter/2-1-1018084.

Geiger, Susi, Debbie Harrison, Hans Kjellberg, and Alexandre Mallard, eds. 2014. *Concerned Markets. Economic Ordering for Multiple Values*. Cheltenham: Edward Elgar Publishing.

Gibson, James J. [1979] 2014. "The Theory of Affordances." In *The People, Place, and Space Reader*, ed. Jen Jack Gieseking and William Mangold, 56–60. New York: Routledge.

Gordon, Colin, Graham Burchell, and Peter Miller, eds. 1991. *The Foucault Effect. Studies in Governmentality: With Two Lectures by and an Interview with Michel Foucault*. Chicago: University of Chicago Press.

Grabowski, Timothy B., and Jonathan H. Grabowski. 2019. "Early Life History." In *Atlantic Cod: A Bio-Ecology*, ed. George A. Rose, 133–168. Chichester: Wiley Blackwell.

Granovetter, Mark. 1985. "Economic Action and Social Structure: The Problem of Embeddedness." *American Journal of Sociology* 91 (3): 481–510.

Greimas, Algirdas Julien, and Joseph Courtés. 1982. *Semiotics and Language. An Analytical Dictionary*. Bloomington: Indiana University Press.

Grytås, Gunnar. 2013. *Motmakt og samfunnsbygger: Med torsken og Norges råfisklag gjennom 75 år* [Counter-power and community builder: With the cod and Norwegian Fishermen's Sales Organization through 75 years]. Trondheim: Akademika forlag.

Gulbrandsen, Magnus. 2011. "Kristian Birkelands spøkelse." In *Universitetet i Oslo 1811–2011. Bok 7: Samtidshistoriske perspektiver*, ed. Peder Anker, Magnus Gulbrandsen, Eirinn Larsen, Johannes W. Løvhaug, and Bent Sofus Tranøy, 275–278. Oslo: Unipub.

Guyer, Jane I. 2009. "Composites, Fictions, and Risk: Toward an Ethnography of Price." In *Market and Society: The Great Transformation Today*, ed. Chris Hann and Keith Hart, 203–220. Cambridge: Cambridge University Press.

Hansen, Tommy. 2020. "Mest torsk på Myre [Most cod on Myre]." *Vesterålen Online*, March 18, 2020. https://www.vol.no/pluss/2020/03/18/Mest-torsk-på-Myre-21372627.ece.

Haraway, Donna J. 1988. "Situated Knowledges: The Science Question in Feminism and the Privilege of Partial Perspective." *Feminist Studies* 14 (3): 575–599.

Haraway, Donna J. 1989. *Primate Visions: Gender, Race, and Nature in the World of Modern Science*. New York: Routledge.

Haraway, Donna J. 1997. *Modest_Witness@Second_Millennium. FemaleMan_Meets_OncoMouse: Feminism and Technoscience*. New York: Routledge.

Haraway, Donna J. 2003. *The Companion Species Manifesto: Dogs, People, and Significant Otherness*. Chicago: Prickly Paradigm Press.

Haraway, Donna J. 2008. *When Species Meet*. Minneapolis: University of Minnesota Press.

Hardin, Garrett. 1968. "The Tragedy of the Commons." *Science* 162 (3859): 1243–1248.

Harvey, David. 2007. *A Brief History of Neoliberalism*. New York: Oxford University Press.

Harvey, David. 2010. *The Enigma of Capital and the Crises of Capitalism*. Oxford: Oxford University Press.

Hastrup, Kirsten, and Frida Hastrup, eds. 2016. *Waterworlds: Anthropology in Fluid Environments*. New York: Berghahn Books.

Helgesson, Claes-Fredrik, and Fabian Muniesa. 2013. "For What It's Worth: An Introduction to Valuation Studies." *Valuation Studies* 1 (1): 1–10.

Helmreich, Stefan. 2008. "Species of Biocapital." *Science as Culture* 17 (4): 463–478.

Helmreich, Stefan. 2009. *Alien Ocean: Anthropological Voyages in Microbial Seas*. Berkeley: University of California Press.

Hersoug, Bjørn. 2005. *Closing the Commons: Norwegian Fisheries from Open Access to Private Property*. Delft: Eburon Uitgeverij BV.

Hersoug, Bjørn, Pål Christensen, and Bjørn-Petter Finstad. n.d. "Råfiskloven—Fra kriseløsning til omstridt monopol [The Raw Fish Act—From crisis resolution to controversial monopoly]." Accessed March 18, 2020. https://www.regjeringen.no/globalassets/upload/fkd/vedlegg/rapporter/2011/vedlegg1_historisk_beskrivelse.pdf.

Herzig, Rebecca, and Banu Subramaniam. 2017. "Labor in the Age of 'Bio-Everything.'" *Radical History Review* 2017 (127): 103–124.

Hetherington, Kregg. 2019. "The Concentration of Killing: Soy, Labor, and the Long Green Revolution." In *How Nature Works: Rethinking Labor on a Troubled Planet*, ed. Sarah Besky and Alex Blanchette, 41–59. Albuquerque: University of New Mexico Press.

Hetherington, Kregg. 2020. *The Government of Beans: Regulating Life in the Age of Monocrops*. Durham, NC: Duke University Press.

Heuts, Frank, and Annemarie Mol. 2013. "What Is a Good Tomato? A Case of Valuing in Practice." *Valuation Studies* 1 (2): 125–146.

Hjort, Jens, J. 1914. *Vekslingerne i de store fiskerier* [The fluctuations of the great fisheries]. Oslo: Aschehoug.

Hobæk, Bård. 2023. "Into the Protein Space." Draft PhD diss., University of Oslo.

Holm, Petter. 1995. "The Dynamics of Institutionalization: Transformation Processes in Norwegian Fisheries." *Administrative Science Quarterly* 40 (3): 398–422.

Holm, Petter. 1996. "Fisheries Management and the Domestication of Nature." *Sociologia Ruralis* 36 (2): 177–188.

Holm, Petter, and Kåre N. Nielsen. 2007. "Framing Fish, Making Markets: The Construction of Individual Transferable Quotas (ITQs)." *The Sociological Review* 55 (2 suppl.): 173–195.

Hommedal, Stine. 2021. "Why Is There So Much Fish on the Fish Banks?" Institute of Marine Research. Last modified July 29, 2021. https://www.hi.no/hi/nyheter/2018/november/hvorfor-er-det-sa-mye-fisk-pa-fiskebankene.

Horowitz, Roger. 2004. "Making the Chicken of Tomorrow: Reworking Poultry as Commodities and as Creatures, 1945–1990." In *Industrializing Organisms: Introducing Evolutionary History*, ed. Susan R. Schrepfer and Philip Scranton, 215–236. London: Routledge.

Hovland, Edgar. 2014. "Havbruksnæringen I krise 1989–1991 [The ocean growing industry in crisis 1989–1991]." In *Over den leiken ville han rå. Norsk havbruksnærings historie* [He wanted to rule breeding. The history of Norwegian ocean growing industry], ed. Edgar Hovland, Dag Møller, Anders Haaland, Nils Kolle, Bjørn Hersoug, and Gunnar Nævdal, 151–177. Bergen: Fagbokforlaget.

Hull, Matthew S. 2012. *Government of Paper*. Berkeley: University of California Press.

Hutchby, Ian. 2001. "Technologies, Texts and Affordances." *Sociology* 35 (2): 441–456.

Hutchinson, Alan. 2014. "Del 5: Lavkonjunktur og nedgangstider i fiskeriene ca. 1600–1720 [Part 5: Recession and period of decline in the fisheries ca. 1600–1720]." In *Norges fiskeri- og kysthistorie bind I: Fangstmenn, fiskerbønder og værfolk, fram til 1720* [Norway's fishery and coastal history vol. I: Hunters, fish farmers and fishing village people], ed. Alf Ragnar Nielssen, 401–546. Oslo: Fagbokforlaget.

Innst. S. Nr. 44. 1965. "Innstilling om utforskning og utnyttelse av undersjøiske petroleumsforekomster med utkast til kongelig resolusjon. Innstilling nr. 2 fra Utvalget til å foreslå regler om utforming av undersjøiske naturforekomster [Recommendation on exploration and exploitation of underwater petroleum deposits with draft for royal resolution. Recommendation nr. 2 from the Committee to propose rules on formulation of underwater nature deposits].

Innst. S. Nr. 327. 1979–1980. "Innstilling fra sjøfarts- og fiskerkomitéen om opprettelse av et olje/fisk fond" (St. prp. nr. 134).

IntraFish. 2018. "Ordforklaringer [Vocabulary]." *IntraFish*, September 22, 2018. https://www.intrafish.no/nyheter/ordforklaringer/2-1-429925.

Jensen, Arne, Sivert Grøntvedt, Harald Skjervold, Ingjald Ørbeck Sørheim, and Eddy G. Torp. 1985. *Å dyrke havet. Perspektivanalyse på norsk havbruk* [Growing the Ocean: Perspectives on Norwegian Ocean Farming]. Trondheim: NTNF and Tapir Forlag. https://urn.nb.no/URN:NBN:no-nb_digibok_2007071300008 (accessed June 30, 2021).

Joint Fish. n.d. "About the Fisheries Commission." Accessed March 21, 2022. https://www.jointfish.com/OM-FISKERIKOMMISJONEN.html.

Kalman Stefánsson, Jón. 2010. *Heaven and Hell.* Translated by Philip Roughton. New York: MacLehose Press.

Karpik, Lucien. 2000. "Le guide rouge Michelin." *Sociologie du Travail* 41: 369–390.

Karpik, Lucien. 2010. *Valuing the Unique: The Economics of Singularities*. Princeton, NJ: Princeton University Press.

Kitchin, Rob, Martin Dodge, and Chris Perkins, eds. 2011. "Conceptualising Mapping." In *The Map Reader: Theories of Mapping Practice and Cartographic Representation*, 1–7. Chichester: Wiley-Blackwell.

Knorr-Cetina, Karin D. 1981. *The Manufacture of Knowledge. An Essay on the Constructivist and Contextual Nature of Science*. Oxford: Pergamon Press.

Knox, Hannah. 2020. *Thinking like a Climate*. Durham, NC: Duke University Press.

Kole, Adriaan, Rian A. A. M. Schelvis-Smit, Martine Veldman, and Joop B. Luten. 2003. *Consumer Perception of Wild and Farmed Cod and the Effect of Different Information Conditions*. RIVO Report C047/03. Netherlands Institute for Fisheries Research (RIVO). http://edepot.wur.nl/151173.

Kole, Adriaan P. W., Themistoklis Altintzoglou, Rian A. A. M. Schelvis-Smit, and Joob B. Luten. 2009. "The Effects of Different Types of Product Information on the Consumer Product Evaluation for Fresh Cod in Real Life Settings." *Food Quality and Preference* 20 (3) (April): 187–194.

Kolle, Nils. 2014. "En næring for distriktene [An industry for the districts]." In *Over den leiken ville han rå. Norsk havbruksnærings historie* [He wanted to rule breeding. The history of Norwegian ocean growing industry], ed. Edgar Hovland, Dag Møller, Anders Haaland, Nils Kolle, Bjørn Hersoug, and Gunnar Nævdal, 151–177. Bergen: Fagbokforlaget.

Kolle, Nils, Alf Ragnar Nielssen, Atle Døssland, and Pål Christensen, eds. 2017. *Fish, Coast and Communities: A History of Norway*. Bergen: Fagbokforlaget.

Krick, Eva, Johan Christensen, and Cathrine Holst. 2019. "Between 'Scientization and a 'Participatory Turn.' Tracing Shifts in the Governance of Policy Advice." *Science and Public Policy* 46 (6) (December): 927–939. https://doi.org/10.1093/scipol/scz040.

Kurlansky, Mark. 1997. *Cod: A Biography of the Fish That Changed the World*. New York: Walker.

Kvenseth, Per Gunnar, and Victor Øiestad. 1984. *Large-Scale Rearing of Cod Fry on the Natural Food Production in Enclosed Pond*. Flødevigen rapporter, l. The Propagation of Cod *Gudus morhua* L.

Laloë, Anne-Flore. 2016. *The Geography of the Ocean: Knowing the Ocean as a Space*. London: Routledge.

Larsen, Per Marius. 1985. "Torskeyngel på skolebenken i Austevoll: Datastyrte anlegg gir god lønnsomhet." *Fiskets Gang nr. 15/16* (32): 539–540. Published by the Director of Fisheries.

Latour, Bruno. 1987. *Science in Action: How to Follow Scientists and Engineers through Society*. Cambridge, MA: Harvard University Press.

Latour, Bruno. 1990. "Technology Is Society Made Durable." *The Sociological Review* 38 (1): 103–131.

Latour, Bruno. 1992. "Drawing Things Together." In *Representation in Scientific Practice*, ed. Michael E. Lynch and Steve Woolgar, 19–68. Cambridge, MA: MIT Press.

Latour, Bruno. 1996. "On Interobjectivity." *Mind, Culture, and Activity* 3 (4): 228–245.

Latour, Bruno. 1998. "To Modernize or to Ecologize? That's the Question." In *Remaking Reality: Nature at the Millenium*, ed. Bruce Braun and Noel Castree, 221–242. London: Routledge.

Latour, Bruno. 2004. *Politics of Nature: How to Bring the Sciences into Democracy*. Cambridge, MA: Harvard University Press.

Latour, Bruno. 2007. "Turning around Politics: A Note on Gerard de Vries' Paper." *Social Studies of Science* 83 (85): 811–820.

Latour, Bruno, and Steve Woolgar. 1979. *Laboratory Life: The Social Construction of Scientific Facts*. Beverly Hills, CA: Sage Publications.

Law, John. 1994. *Organizing Modernity*. Oxford: Blackwell.

Law, John. 2004. *After Method: Mess in Social Science Research*. London: Routledge.

Law, John. 2009. "Seeing like a Survey." *Cultural Sociology* 3 (2): 239–256.

Law, John. 2010a. "Care and Killing: Tensions in Veterinary Practice." In *Care in Practice: On Tinkering in Clinics, Homes and Farms*, ed. Annemarie Mol, Ingunn Moser, and Jeannette Pols, 57–69. Bielefield: Transcript Publishers.

Law, John. 2010b. "The Double Social Life of Method." Presentation prepared for the Sixth Annual CRESC conference on the Social Life of Method, August 31–September 3, St. Hugh's College, Oxford. http://w.heterogeneities.net/publications/Law2010DoubleSocialLifeofMethod5.pdf.

Law, John, Geir Afdal, Kristin Asdal, Wen-yuan Lin, Ingunn Moser, and Vicky Singleton. 2014. "Modes of Syncretism: Notes on Noncoherence." *Common Knowledge* 20 (1): 172–192.

Law, John, and John Hassard. 1999. *Actor Network Theory and After*. Oxford: Wiley-Blackwell.

Law, John, and Marianne E. Lien. 2013. "Slippery: Field Notes in Empirical Ontology." *Social Studies of Science* 43 (3): 363–378.

Law, John, and Annemarie Mol. 2020. "Words to Think With: An Introduction." *Sociological Review* 68 (2): 263–282.

Law, John, and Evelyn Ruppert. 2013. "The Social Life of Methods: Devices." *Journal of Cultural Economy* 6 (3): 229–240.

Lien, Marianne E. 2015. *Becoming Salmon: Aquaculture and the Domestication of a Fish*. Oakland: University of California Press.

Lien, Marianne E., Heather A. Swanson, and Gro Birgit Ween. 2018. "Introduction: Naming the Beast—Exploring the Otherwise." In *Domestication Gone Wild*, 1–30. Durham, NC: Duke University Press.

Luten, Joop, Adriaan Kole, Rian Schelvis, Martine Veldman, Morten Heide, Mats Carlehög, and Leif Akse. 2002. "Evaluation of Wild Cod versus Wild Caught, Farmed Raised Cod from Norway by Dutch Consumers." *Økonomisk Fiskeriforskning* 12: 44–60.

Lysvold, Susanne Skjåstad. 2019. "Norwegian Cod Is Sent to China and Filled with Water and Chemicals." *NRK*, January 23, 2019. https://www.nrk.no/nordland/norsk-torsk-sendes-til-kina-og-fylles-med-vann-og-kjemikalier-1.14393315.

MacKenzie, Donald. 2006. *An Engine, Not a Camera: How Financial Models Shape Markets*. Cambridge, MA: MIT Press.

MacKenzie, Donald, Fabian Muniesa, and Lucia Siu, eds. 2007. *Do Economists Make Markets? On the Performativity of Economics*. Princeton, NJ: Princeton University Press.

Malinowski, Bronisław. 1922. *Argonauts of the Western Pacific*. London: Routledge.

Mandeville, Bernard. (1924) 1988. *The Fable of the Bees; or, Private Vices, Publick Benefits*. Original manuscript from 1732. Altenmünster: Jazzybee Verlag.

Marinbiolegene DA [The Marine Biologists]. 2016. "Skreien den glade vandrar [Skrei, the happy wanderer]." Last modified March 3, 2016. http://marinbiologene.no/skreien-den-glade-vandrar/.

Mariussen, Åge. 1992. *Fra vann til hav. Prosessevaluering av havbruk som forskningspolitisk hovedinnsatsområde* [From water to ocean. Process evaluation of ocean growing as political research priority area]. Norwegian Fisheries Research Council. https://urn.nb.no/URN:NBN:no-nb_digibok_2015061808082.

Marx, Karl. (1867) 2018. *Capital: A Critique of Political Economy*. Volume I. Book One: The Process of Production of Capital. Mission, BC: Modern Barbarian Press.

Mauss, Marcel. 1954. *The Gift: Forms and Functions of Exchange in Archaic Society*. London: Cohen & West.

McKinsey & Company. 2013. "China's Next Chapter." *McKinsey Quarterly 2013* (3). https://mckinsey.com/~/media/mckinsey/mckinsey%20quarterly/digital%20newsstand/2013%20issues%20mckinsey%20quarterly/chinas%20next%20chapter.pdf (accessed June 26, 2021).

Meadows, Donella H., Dennis L. Meadows, Jørgen Randers, and William W. Behrens III. 1972. *The Limits to Growth: A Report for the Club of Rome's Project on the Predicament of Mankind*. New York: Universe Books.

Midling, Kjell Ø. 1990. "Fjordbeite med kondisjonert torsk [Fjord pasture with conditioned cod]." Master's thesis, University of Bergen.

Midling, Kjell Ø., Tore S. Kristiansen, Egil Ona, and Victor Øiestad. 1987. "Fjord-ranching with Conditioned Cod." *International Council for the Exploration of the Sea.* Mariculture Committee.

Moksness, Erlend, and Victor Øiestad. 1984. "Tagging and Release Experiments on 0-Group Coastal Cod (*Gadus morhua* L.) Reared in an Outdoor Basin." *Flødevigen rapportser,* 1. The Propagation of Cod *Gadus morhua* L.

Mol, Annemarie. 1999. "Ontological Politics: A Word and Some Questions." In *Actor Network Theory and After,* ed. John Law and John Hassard, 74–89. Oxford: Wiley-Blackwell.

Mol, Annemarie. 2002. *The Body Multiple.* Durham, NC: Duke University Press.

Moore, Jason. 2015. *Capitalism in the Web of Life: Ecology and the Accumulation of Capital.* New York: Verso Books.

Moser, Ingunn. 2008. "Making Alzheimer's Disease Matter: Enacting, Interfering and Doing Politics of Nature." *Geoforum* 39 (1): 98–110.

Muniesa, Fabian. 2011. "A Flank Movement in the Understanding of Valuation." *Sociological Review* 59 (2 suppl.): 24–38.

Muniesa, Fabian. 2014. *The Provoked Economy: Economic Reality and the Performative Turn.* London: Routledge.

Muniesa, Fabian, Liliana Doganova, Horacio Ortiz, Álvaro Pina-Stranger, Florence Paterson, Alaric Bourgoin, Véra Ehrenstein, Pierre-André Juven, David Pontille, Başak Saraç-Lesavre, and Guillaume Yon. 2017. *Capitalization: A Cultural Guide.* Paris: Presses des Mines.

Muniesa, Fabian, Yuval Millo, and Michel Callon. 2007. "An Introduction to Market Devices." *Sociological Review* 55 (2): 1–12.

Murphy, Michelle. 2017. *The Economization of Life.* Durham, NC: Duke University Press.

Musselin, Christine, and Catherine Paradeise. 2005. "Quality: A Debate." *Sociologie du Travail* 47: e89–e123.

Nævdal, Gunnar, and Edgar Hovland. 2014. "Lang dags flrd—Hvor hen? Marine arter i norsk havbruk [Long day's journey—Where to? Marine species in Norwegian ocean growing]." In *Over den leiken ville han rå. Norsk havbruksnærings historie* [He wanted to rule breeding. The history of Norwegian ocean growing industry], ed. Edgar Hovland, Dag Møller, Anders Haaland, Nils Kolle, Bjørn Hersoug, and Gunnar Nævdal, 353–390. Bergen: Fagbokforlaget.

Nebdal, Sofie. 2019. "Å gjøre havet økonomisk: Verdsettingspraksiser for økonomisering av havet og dets økosystemer." Master's thesis, University of Oslo.

National Museums Scotland. n.d. "Dolly the Sheep." Accessed March 4, 2022. https://www.nms.ac.uk/explore-our-collections/stories/natural-sciences/dolly-the-sheep/.

Nicholas, Howard. 2012. "What Is the Problem with Neoclassical Price Theory?" *World Review of Political Economy* 3 (4): 457–477. Doi:10.13169/worlrevipoliecon.3.4.0457.

Nielsen, Reidar. 2012. "Havets gull: Fiskekjøpere i Nord-Troms og Finnmark [Gold of the sea: Fish buyers in Nord-Troms and Finnmark]." Fiskeprodusentenes fond.

Nik-Khah, Edward, and Philip Mirowski. 2019. "One Going the Market One Better: Economic Market Design and the Contradictions of Building Markets for Public Purposes." *Economy and Society* 48 (2): 268–294.

Nodland, Elizabeth. 2016. "Marine Harvest Is Suing Norway for ESA." *iLaks.no*, May 2, 2016. https://ilaks.no/marine-harvest-klager-norge-inn-for-esa/.

Norberg, Birgitta. 2002. "Project Description: Cod Puberty—Age at First Maturation in Relation to Season, Growth End Energy Acquisition During the First Year of Life." Forsiden—Norges Forskningsråd. https://www.forskningsradet.no/prosjektbanken_beta/ #/project/146677.

NOU 1977. *Fiskeoppdrett* [Fish farming]. Oslo: Ministry of Fisheries.

NOU 1978. *Olje- og fiskerinæringen*. Oslo: Ministry of Oil and Petroleum.

NOU 1985. *Akvakultur i Norge. Status og framtidsutsikter* [Aquaculture in Norway: Status and prospects for the future]. Oslo: Ministry of Fisheries.

NOU 1986. *Erstatning til fiskerne for ulemper ved petroleumsvirksomheten*. Oslo: Ministry of Oil and Petroleum.

NOU 2000. *Ny giv for nyskaping* [A new go for value creation]. Oslo: Ministry of Business and Commerce. https://www.regjeringen.no/no/dokumenter/nou-2000-7/id376058/?ch=1.

OECD. 2016. *The Ocean Economy in 2030*. Paris: OECD Publishing. https://doi.org/10.1787/9789264251724-en.

Office of the Auditor General of Norway. 2020. *Unlersøkelse av kvotesystemet i kyst- og havfisket*. Document 3: 6 (2019–2020). https://www.riksrevisjonen.no/globalassets/rapporter/no-2019-2020/kvotesystemet-i-kyst--og-havfisket.pdf.

Øiestad, Victor. 1985. "Produksjon av torskeyngel i poll [Production of cod spawn in pool]." In *Olje/Fisk Fondet* 1980–1985.

Øiestad, Victor. 1990. *Torsk i oppdrelt gjennom hundre år. Håndbok i torskeoppdrett. Stamfiskhold og yngelproduksjon* [100 years of farmed cod. Handbook in cod farming. Broodstock and spawn production]. Bergen: Institute of Marine Research.

Øiestad, Victor. 1994. "Fjordbeide med dressert torsk—Fiskeren som ressursvennlig bonde [Fjord pasture with trained cod—The fisher as resource friendly farmer]." *Fiskets gang, nr.* (2): 19–23.

Øiestad, Victor, Torstein Pedersen, Arild Folkvord, Åsmund Bjordal, and Per G. Kvenseth. 1987. "Automatic Feeding and Harvesting of Juvenile Atlantic Cod in a Pond." *Modeling, Identification and Control* 8 (1): 39–46.

Online Etymology Dictionary. n.d. *dem-. Accessed June 14, 2021. https://www.etymonline.com/word/*dem-.

Online Etymology Dictionary. n.d. "domesticate" (v.). Accessed June 14, 2021. https://www.etymonline.com/word/domesticate.

O'Sullivan, Mary. 2005. "Finance and Innovation." In *The Oxford Handbook of Innovation*, ed. Jan Fagerberg, David Mowery, and Richard Nelson, 240–265. Oxford: Oxford University Press.

Pallesen, Trine. 2016. "Valuation Struggles over Pricing—Determining the Worth of Wind Power." *Journal of Cultural Economy* 9 (6): 1–14.

Parliamentary Proceedings. 1985–86. Bev. på statsbudsj. 1986 vedk. Handelsdep. og Fiskeridep. [Grants on the National Budget Concerning the Ministry of Trade and the Ministry of Fisheries]. In Stortingsforhandlinger 1985–86 [Parliamentary Proceedings 1985–86] Vol. 130 Nr. 7a, 620–671.

Peck, Jamie. 2010. *Constructions of Neoliberal Reason*. Oxford: Oxford University Press.

Polanyi, Karl. (1944) 2001. *The Great Transformation: The Political and Economic Origins of Our Time*. New York: Farrar & Rinehart.

Rajan, Kaushik S. 2006. *Biocapital: The Constitution of Postgenomic Life*. Durham, NC: Duke University Press.

Rajan, Kaushik S. 2012. "Introduction: The Capitalization of Life and the Liveliness of Capital." In *Lively Capital: Biotechnologies, Ethics, and Governance in Global Markets*, ed. Rajan S. Kaushik, 1–44. Durham, NC: Duke University Press.

Redmayne, Peter, and Sea Fare Expositions. 2018. *Greetings from the Overseas Organizer*. Expo handbook.

Reinecke, Juliane 2010. "Beyond a Subjective Theory of Value and towards a 'Fair Price': An Organizational Perspective on Fairtrade Minimum Price Setting." *Organization* 17 (5): 563–581.

Reinertsen, Hilde, and Kristin Asdal. 2019. "Calculating the Blue Economy: Producing Trust in Numbers with Business Tools and Reflexive Objectivity." *Journal of Cultural Economy* 12 (6): 552–570. https://doi.org/10.1080/17530350.2019.1639066.

Research Council of Norway and Innovation Norway (RCN/IN). 2006. *Plan for koordinert satsing på torsk. Oppdrett og fangstbasert akvakultur. 2001–2010* [Plan for a Coordinated Investment in Cod: Farming and Capture-Based Aquaculture. 2001–2010]. Updated April 2006.

Research Council of Norway, Innovation Norway, and Norwegian Seafood Research Fund (RCN/IN/NSRF). 2009. *Plan for koordinert satsing på torsk. Oppdrett og fangstbasert akvakultur. 2001–2010* [Plan for a Coordinated Investment in Cod: Farming and Capture-Based Aquaculture. 2001–2010]. Updated May 2009.

Research Council of Norway and the State's Industry and District Development Fund (RCN/SND). 2001. *Oppdrett av torsk. Strategi for koordinert satsing fra SND og Norges forskningsråd 2001–2010* [Farming of Cod: Strategy for Coordinated Effort by SND and the Research Council of Norway 2001–2010].

Research Council of Norway and the State's Industry and District Development Fund (RCN/SND). 2003. *Oppdrett av torsk. Strategi for koordinert satsing fra SND og Norges forskningsråd 2001–2010* [Farming of Cod: Strategy for Coordinated Effort by SND and the Research Council of Norway 2001–2010]. Updated July 2003.

Ricardo, David. 1817. *On the Principles of Political Economy and Taxation*. London: John Murray.

Roorda, Eric P. 2020. "Introduction." In *The Ocean Reader: History, Culture, Politics*, ed. Eric P. Roorda, 1–4. Durham, NC: Duke University Press.

Rose, George A., ed. 2019. *Atlantic Cod: A Bio-Ecology*. Chichester: Wiley-Blackwell.

Rose, Nikolas. 2007. "Molecular Biopolitics, Somatic Ethics and the Spirit of Biocapital." *Social Theory & Health* 5 (1): 3–29.

Rose-Redwood, Reuben, and Liora Bigon, eds. 2018. *Gridded Worlds: An Urban Anthology*. Cham: Springer International Publishing.

Ruud, Morten, and Geir Ulfstein. 2011. *Innføring i folkerett* [Introduction to international law]. 4th ed. Oslo: Universitetsforlaget.

SalMar ASA. 2019. "Ocean Farm 1." Facebook, February 14, 2019. https://www.facebook.com/watch/?v=324719204842322.

Sars, Georg Ossian. 1869. *Indberetninger til Departementet for det Indre fra G. O. Sars, om de af ham i Aarene 1864–69 anstillede praktisk-videnskabelige Undersøgelser angaaende Torskefiskeriet i Lofoten* [Report to Ministry of the Interior from G. O. Sars on his practical scientific studies during the years 1864–1869 regarding cod fishing in Lofoten]. Christiania: Det Steenske bogtrykkeri.

Savage, Mike. 2013. "The 'Social Life of Methods': A Critical Introduction." *Theory, Culture & Society* 30 (4): 3–21.

Schabas, Margaret. 2005. *The Natural Origins of Economics*. Chicago: University of Chicago Press.

Schlünder, Martina. 2017. "The Measure of the Disease: The Pathological Animal Experiment in Robert Koch's Medical Bacteriology." In *Humans, Animals and Biopolitics: The More-Than-Human Condition*, ed. Kristin Asdal, Tone Druglitrø, and Steve Hinchliffe, 101–120. Abingdon: Routledge.

Schumacher, Ernst Friedrich. 1973. *Small Is Beautiful*. New York: Harper and Row.

Schumpeter, Joseph. A. (1934) 2008. *The Theory of Economic Development*. New Brunswick, NJ: Transaction Publishers.

Schwach, Vera. 2013. "The Sea around Norway: Science, Resource Management, and Environmental Concerns, 1860–1970." *Environmental History* 18 (1): 101–110.

Skinner, Quentin. 1969. "Meaning and Understanding in the History of Ideas." *History and Theory* 8 (1): 3–53.

Smith, Adam. (1776) 2008. *An Inquiry into the Nature and Causes of the Wealth of Nations*. A selected edition. Edited by K. Sutherland. Oxford: Oxford University Press.

Soluri, John. 2002. "Accounting for Taste: Export Bananas, Mass Markets, and Panama Disease." *Environmental History* 7 (3): 386–410.

Soluri, John. 2005. *Banana Cultures: Agriculture, Consumption, and Environmental Change in Honduras and the United States*. Austin: University of Texas Press.

Stanghelle, Harald. 2017. "Commentary: The Silent Minister of Foreign Affairs." *Aftenposten*, May 29, 2017. https://www.aftenposten.no/meninger/i/GWW3m/kommentar-den-tause-ytringsministeren.

Star, Bastiaan, Sanne Boessenkool, Agata T. Gondek, Elena A. Nikulina, Anne Karin Hufthammer, Cristophe Pampoulie, Halvor Knutsen, Carl André, Heidi M. Nistelberger, Jan Dierking, Christoph Petereit, Dirk Heinrich, Kjetill S. Jakobsen, Nils Chr. Stenseth, Sissel Jentoft, and James H. Barrett. 2017. "Ancient DNA Reveals the Arctic Origin of Viking Age Cod from Haithabu, Germany." *Proceedings of the National Academy of Sciences* 114 (34): 9152–9157. https://doi.org/10.1073/pnas.1710186114.

Stark, David. 2011a. *The Sense of Dissonance: Accounts of Worth in Economic Life*. Princeton, NJ: Princeton University Press.

Stark, David. 2011b. "Postscript." In *The Worth of Goods: Valuation and Pricing in the Economy*, ed. Jens Beckert and Patrik Aspers, 319–338. Oxford: Oxford University Press.

Statt Torsk. n.d. "A New Frontier for Norwegian Aquaculture." Accessed June 3, 2020. https://statt.no/en/.

Steinberg, Philip E. 2001. *The Social Construction of the Ocean*. Cambridge: Cambridge University Press.

Steinberg, Philip E., and Berit Kristoffersen. 2017. "'The ice edge is lost . . . nature moved it': Mapping Ice as State Practice in the Canadian and Norwegian North." *Transactions of the Institute of British Geographers* 42 (4): 625–641.

St. meld. 1978–1979. *Petroleumsundersøkelser nord for 62°N* [Petroleum explorations north of 62°N]. No. 57. Oslo: Ministry of Oil and Energy.

St. meld. 1986–1987. *Om havbruk* [On ocean growing]. No. 65. Ministry of Fisheries.

St. meld. 2015–2016. *En konkurransekraftig sjømatindustri* [A competitive seafood industry]. No. 10. Ministry of Trade, Industry and Fisheries. https://regjeringen.no/no/dokumenter/meld.-st.-10-20152016/id2461010/ (accessed June 30, 2021).

Store Norske Leksikon [Great Norwegian Encyclopedia]. n.d.a. "The Cod War." Accessed July 30, 2021. https://snl.no/Torskekrigen.

Store Norske Leksikon [Great Norwegian Encyclopedia]. n.d.b. "Gunder Mathisen Dannevig." Accessed June 21, 2021. https://nbl.snl.no/Gunder_Mathiesen_Dannevig.

St. prp. 1979–1980. *Om opprettelse av et olje-fisk fond* [On the establishment of an Oil-Fish Fund]. No. 134. Recommendation to the Ministry of Fisheries, May 30, 1980. Oslo: Ministry of Fisheries.

St. prp. 2002–2003. Virkemidler for et innovativt og nyskapende næringsliv [Measures for an Innovative and Inventive Business Sector]. No. 51. Ministry of Business and Commerce.

Svaar, Peter, Signe Karin Hotvedt, and Kristine Hirsti. 2017. "Chinese Dissident Upset: Is a Fishmonger or a Prime Minister Coming to Visit?" *NRK*, April 3, 2017. https://www.nrk.no/urix/kinesisk-dissident-opprort-over-erna-solberg-1.13456934.

Svendsen, Mette N. 2020. "Pig-Human Relations in Neonatology: Knowing and Unknowing in a Multi-Species Collaborative." In *Biosocial Worlds. Anthropology of Health Environments beyond Determinism*, ed. Jens Seeberg, Andreas Ropestorff, and Lotte Meinert, 69–90. London: UCL Press.

Taranger, Geir Lasse, Leiv Aardal, Tom Hansen, and Olav Sigurd Kjesbu. 2006. "Continuous Light Delays Sexual Maturation and Increases Growth of Atlantic Cod (*Gadus morhua* L.) in Sea Cages." *ICES Journal of Marine Science* 63 (2): 365–375.

Taranger, Geir Lasse, Manuel Carrillo, Rüdiger W. Schulz, Pascal Fontaine, Silvia Zanuy, Alicia Felip, Finn-Arne Weltzien, Sylvie Dufour, Ørjan Karlsen, Birgitta Norberg, Eva Andersson, and Tom Hansen. 2010. "Control of Puberty in Farmed Fish." *General and Comparative Endocrinology* 165 (3): 483–515.

Teigen, Håvard. 2019. *Distriktspolitikkens historie i Norge* [The history of regional politics in Norway]. Oslo: Cappelen Damm Akademisk.

Telesca, Jennifer E. 2020. *Red Gold: The Managed Extinction of the Giant Bluefin Tuna*. Minneapolis: University of Minnesota Press.

Thompson, Charis. 2000. "The Biotech Mode of Reproduction." Paper presented at the School of American Research Advanced Seminar, "Animation and Cessation: Anthropological Perspectives on Changing Definitions of Life and Death in the Context of Biomedicine," Santa Fe, NM.

Thompson, Charis. 2005. *Making Parents: The Ontological Choreography of Reproductive Technologies*. Cambridge, MA: MIT Press.

Thompson, Charis. 2013. *Good Science: The Ethical Choreography of Stem Cell Research*. Cambridge, MA: MIT Press.

Thompson, Edward Palmer. 1971. "The Moral Economy of the English Crowd in the Eighteenth Century." *Past & Present* 50: 76–136.

Trébuchet-Breitwiller, Anne-Sophie. 2015. "Makings Things Precious: A Pragmatist Inquiry into the Valuation of Luxury Perfumes." In *Moments of Valuation: Exploring*

Sites of Dissonance, ed. Ariane Berthoin Antal, Michael Hutter, and David Stark, 168–186. Oxford: Oxford University Press.

Treves, Trulio. 2008. "1958 Geneva Conventions on the Law of the Sea." Audiovisual Library of International Law. Last modified September 2008. https://legal.un.org/avl/ha/gclos/gclos.html.

Tsing, Anna L. 2015. *The Mushroom at the End of the World*. Princeton, NJ: Princeton University Press.

UN. 1982. United Nations Convention on the Law of the Sea.

Vatin, François. 2013. "Valuation as Evaluating and Valorizing." *Valuation Studies* 1 (1): 31–50.

Vora, Kalindi. 2015. *Life Support: Biocapital and the New History of Outsourced Labor*. Minneapolis: University of Minnesota Press.

Voyer, Michelle, Genevieve Quirk, Alistair McIlgorm, and Kamal Azmi. 2018. "Shades of Blue: What Do Competing Interpretations of the Blue Economy Mean for Oceans Governance?" *Journal of Environmental Policy & Planning* 20 (5): 595–616.

Weber, Max. (1904) 2001. *The Protestant Ethic and the Spirit of Capitalism*. Translated by Talcott Parsons. London: Routledge.

Whatmore, Sarah. 2002. *Hybrid Geographies: Natures Cultures Spaces*. London: Sage Publishing.

Wigen, Kären. 2006. "Introduction." *American Historical Review* 111 (3): 717–721.

Wilkie, Rhoda. 2010. *Livestock/Deadstock: Working with Farm Animals from Birth to Slaughter*. Philadelphia: Temple University Press.

Yan, Yunxiang. 2013. "The Gift and Gift Economy." In *A Handbook of Economic Anthropology*. 2nd ed., ed. James G. Carrier, 275–290. Cheltenham: Edward Elgar Publishing.

Yoxen, Edward. 1981. "Life as a Productive Force: Capitalising the Science and Technology of Molecular Biology." In *Science, Technology and the Labour Process*, ed. Les Levidow and Robert M. Young, 66–122. London: Blackrose Press.

Zelizer, Viviana. 1979. *Morals and Markets: The Development of Life Insurance in the United States*. New York: Columbia University Press.

Zelizer, Viviana. 1985. *Pricing the Priceless Child: The Changing Social Value of Children*. New York: Basic Books.

INDEX

Note: References to legislation—for example, Foreign Service Act—are to Norwegian legislation unless otherwise noted.

Inside Technology Series

Edited by Wiebe E. Bijker and Rebecca Slayton

Kristin Asdal and Tone Huse, *Nature-Made Economy: Cod, Capital, and the Great Economization of the Ocean*

Cyrus C. M. Mody, *The Squares: US Physical and Engineering Scientists in the Long 1970s*

Brice Laurent, *European Objects: The Troubled Dreams of Harmonization*

Florian Jaton, *The Constitution of Algorithms: Ground-Truthing, Programming, Formulating*

Kean Birch and Fabian Muniesa, *Turning Things into Assets*

David Demortain, *The Science of Bureaucracy: Risk Decision-Making and the US Environmental Protection Agency*

Nancy Campbell, *OD: Naloxone and the Politics of Overdose*

Lukas Engelmann and Christos Lynteris, *Sulphuric Utopias: The History of Maritime Fumigation*

Zara Mirmalek, *Making Time on Mars*

Joeri Bruynincx, *Listening in the Field: Recording and the Science of Birdsong*

Edward Jones-Imhotep, *The Unreliable Nation: Hostile Nature and Technological Failure in the Cold War*

Jennifer L. Lieberman, *Power Lines: Electricity in American Life and Letters, 1882–1952*

Jess Bier, *Mapping Israel, Mapping Palestine: Occupied Landscapes of International Technoscience*

Benoît Godin, *Models of Innovation: The History of an Idea*

Stephen Hilgartner, *Reordering Life: Knowledge and Control in the Genomics Revolution*

Brice Laurent, *Democratic Experiments: Problematizing Nanotechnology and Democracy in Europe and the United States*

Cyrus C. M. Mody, *The Long Arm of Moore's Law: Microelectronics and American Science*

Tiago Saraiva, *Fascist Pigs: Technoscientific Organisms and the History of Fascism*

Teun Zuiderent-Jerak, *Situated Interventions: Sociological Experiments in Healthcare*

Basile Zimmermann, *Technology and Cultural Difference: Electronic Music Devices, Social Networking Sites, and Computer Encodings in Contemporary China*

Andrew J. Nelson, *The Sound of Innovation: Stanford and the Computer Music Revolution*

Sonja D. Schmid, *Producing Power: The Pre-Chernobyl History of the Soviet Nuclear Industry*

Casey O'Donnell, *Developer's Dilemma: The Secret World of Videogame Creators*

Christina Dunbar-Hester, *Low Power to the People: Pirates, Protest, and Politics in FM Radio Activism*

Eden Medina, Ivan da Costa Marques, and Christina Holmes, editors, *Beyond Imported Magic: Essays on Science, Technology, and Society in Latin America*

Anique Hommels, Jessica Mesman, and Wiebe E. Bijker, editors, *Vulnerability in Technological Cultures: New Directions in Research and Governance*

Amit Prasad, *Imperial Technoscience: Transnational Histories of MRI in the United States, Britain, and India*

Charis Thompson, *Good Science: The Ethical Choreography of Stem Cell Research*

Tarleton Gillespie, Pablo J. Boczkowski, and Kirsten A. Foot, editors, *Media Technologies: Essays on Communication, Materiality, and Society*

Catelijne Coopmans, Janet Vertesi, Michael Lynch, and Steve Woolgar, editors, *Representation in Scientific Practice Revisited*

Rebecca Slayton, *Arguments That Count: Physics, Computing, and Missile Defense, 1949–2012*

Stathis Arapostathis and Graeme Gooday, *Patently Contestable: Electrical Technologies and Inventor Identities on Trial in Britain*

Jens Lachmund, *Greening Berlin: The Co-Production of Science, Politics, and Urban Nature*

Chikako Takeshita, *The Global Biopolitics of the IUD: How Science Constructs Contraceptive Users and Women's Bodies*

Cyrus C. M. Mody, *Instrumental Community: Probe Microscopy and the Path to Nanotechnology*

Morana Alač, *Handling Digital Brains: A Laboratory Study of Multimodal Semiotic Interaction in the Age of Computers*

Gabrielle Hecht, editor, *Entangled Geographies: Empire and Technopolitics in the Global Cold War*

Michael E. Gorman, editor, *Trading Zones and Interactional Expertise: Creating New Kinds of Collaboration*

Matthias Gross, *Ignorance and Surprise: Science, Society, and Ecological Design*

Andrew Feenberg, *Between Reason and Experience: Essays in Technology and Modernity*

Wiebe E. Bijker, Roland Bal, and Ruud Hendricks, *The Paradox of Scientific Authority: The Role of Scientific Advice in Democracies*

Park Doing, *Velvet Revolution at the Synchrotron: Biology, Physics, and Change in Science*

Gabrielle Hecht, *The Radiance of France: Nuclear Power and National Identity after World War II*

Richard Rottenburg, *Far-Fetched Facts: A Parable of Development Aid*

Michel Callon, Pierre Lascoumes, and Yannick Barthe, *Acting in an Uncertain World: An Essay on Technical Democracy*

Ruth Oldenziel and Karin Zachmann, editors, *Cold War Kitchen: Americanization, Technology, and European Users*

Deborah G. Johnson and Jameson W. Wetmore, editors, *Technology and Society: Building Our Sociotechnical Future*

Trevor Pinch and Richard Swedberg, editors, *Living in a Material World: Economic Sociology Meets Science and Technology Studies*

Christopher R. Henke, *Cultivating Science, Harvesting Power: Science and Industrial Agriculture in California*

Helga Nowotny, *Insatiable Curiosity: Innovation in a Fragile Future*

Karin Bijsterveld, *Mechanical Sound: Technology, Culture, and Public Problems of Noise in the Twentieth Century*

Peter D. Norton, *Fighting Traffic: The Dawn of the Motor Age in the American City*

Joshua M. Greenberg, *From Betamax to Blockbuster: Video Stores and the Invention of Movies on Video*

Mikael Hård and Thomas J. Misa, editors, *Urban Machinery: Inside Modern European Cities*

Christine Hine, *Systematics as Cyberscience: Computers, Change, and Continuity in Science*

Wesley Shrum, Joel Genuth, and Ivan Chompalov, *Structures of Scientific Collaboration*

Shobita Parthasarathy, *Building Genetic Medicine: Breast Cancer, Technology, and the Comparative Politics of Health Care*

Kristen Haring, *Ham Radio's Technical Culture*

Atsushi Akera, *Calculating a Natural World: Scientists, Engineers and Computers during the Rise of U.S. Cold War Research*

Donald MacKenzie, *An Engine, Not a Camera: How Financial Models Shape Markets*

Geoffrey C. Bowker, *Memory Practices in the Sciences*

Christophe Lécuyer, *Making Silicon Valley: Innovation and the Growth of High Tech, 1930–1970*

Anique Hommels, *Unbuilding Cities: Obduracy in Urban Sociotechnical Change*

David Kaiser, editor, *Pedagogy and the Practice of Science: Historical and Contemporary Perspectives*

Charis Thompson, *Making Parents: The Ontological Choreography of Reproductive Technologies*

Pablo J. Boczkowski, *Digitizing the News: Innovation in Online Newspapers*

Dominique Vinck, editor, *Everyday Engineering: An Ethnography of Design and Innovation*

Nelly Oudshoorn and Trevor Pinch, editors, *How Users Matter: The Co-Construction of Users and Technology*

Peter Keating and Alberto Cambrosio, *Biomedical Platforms: Realigning the Normal and the Pathological in Late-Twentieth-Century Medicine*

Paul Rosen, *Framing Production: Technology, Culture, and Change in the British Bicycle Industry*

Maggie Mort, *Building the Trident Network: A Study of the Enrollment of People, Knowledge, and Machines*

Donald MacKenzie, *Mechanizing Proof: Computing, Risk, and Trust*

Geoffrey C. Bowker and Susan Leigh Star, *Sorting Things Out: Classification and Its Consequences*

Charles Bazerman, *The Languages of Edison's Light*

Janet Abbate, *Inventing the Internet*

Herbert Gottweis, *Governing Molecules: The Discursive Politics of Genetic Engineering in Europe and the United States*

Kathryn Henderson, *On Line and On Paper: Visual Representation, Visual Culture, and Computer Graphics in Design Engineering*

Susanne K. Schmidt and Raymund Werle, *Coordinating Technology: Studies in the International Standardization of Telecommunications*

Marc Berg, *Rationalizing Medical Work: Decision Support Techniques and Medical Practices*

Eda Kranakis, *Constructing a Bridge: An Exploration of Engineering Culture, Design, and Research in Nineteenth-Century France and America*

Paul N. Edwards, *The Closed World: Computers and the Politics of Discourse in Cold War America*

Donald MacKenzie, *Knowing Machines: Essays on Technical Change*

Wiebe E. Bijker, *Of Bicycles, Bakelites, and Bulbs: Toward a Theory of Sociotechnical Change*

Louis L. Bucciarelli, *Designing Engineers*

Geoffrey C. Bowker, *Science on the Run: Information Management and Industrial Geophysics at Schlumberger, 1920–1940*

Wiebe E. Bijker and John Law, editors, *Shaping Technology / Building Society: Studies in Sociotechnical Change*

Stuart Blume, *Insight and Industry: On the Dynamics of Technological Change in Medicine*

Donald MacKenzie, *Inventing Accuracy: A Historical Sociology of Nuclear Missile Guidance*

Pamela E. Mack, *Viewing the Earth: The Social Construction of the Landsat Satellite System*

H. M. Collins, *Artificial Experts: Social Knowledge and Intelligent Machines*

http://mitpress.mit.edu/books/series/inside-technology